Y0-BRJ-192

Environmental Science

Series Editors: R. Allan • U. Förstner • W. Salomons

Springer
Berlin
Heidelberg
New York
Hong Kong
London
Milan
Paris
Tokyo

TD
398
R55

Piero Risoluti

Nuclear Waste

A Technological and Political Challenge

Springer

Dr. Piero Risoluti
Via Roma 57
00066 Manziana (Rome)

Italy
piero.risoluti@libero.it

This book has been translated from Italian by Piero Risoluti and Mario Guidotti. The original title is as follows:
"I rifiuti nucleari: sfida tecnologica o politica?"

ISBN 3-540-40447-3 Springer-Verlag Berlin Heidelberg New York

Cataloging-in-Publication Data applied for

A catalog record for this book is available from the Library of Congress.

Bibliographic information published by Die Deutsche Bibliothek
Die Deutsche Bibliothek lists this publication in the Deutsche Nationalbibliografie; detailed bibliographic data is available in the Internet at <http://dnb.ddb.de>.

This work is subject to copyright. All rights are reserved, whether the whole or part of the material is concerned, specifically the rights of translation, reprinting, reuse of illustrations, recitation, broadcasting, reproduction on microfilm or in any other way, and storage in data banks. Duplication of this publication or parts thereof is permitted only under the provisions of the German Copyright Law of September 9, 1965, in its current version, and permission for use must always be obtained from Springer-Verlag. Violations are liable for prosecution under the German Copyright Law.

Springer-Verlag Berlin Heidelberg New York
a part of Springer Science+Business Media GmbH

springeronline.com

© Springer-Verlag Berlin Heidelberg 2004
Printed in Germany

The use of general descriptive names, registered names, trademarks, etc. in this publication does not imply, even in the absence of a specific statement, that such names are exempt from the relevant protective laws and regulations and therefore free for general use.

Product liability: The publishers cannot guarantee the accuracy of any information about the application of operative techniques and medications contained in this book. In every individual case the user must check such information by consulting the relevant literature.

Camera ready by the author
Cover design: Struve & Partner, Heidelberg
Printed on acid-free paper 30/2132/AO 5 4 3 2 1 0

Foreword

Over the past decades, Piero Risoluti has built up an intimate knowledge of the nuclear industry - in particular of nuclear waste management. In this book, his scientific understanding is apparent - for example in his comprehensive but readily understandable descriptions of waste conditioning and disposal. Moreover, he has also been directly involved in the wider societal and political debates in the nuclear area - especially in his Italian homeland. What shines through in these pages is his frustration at the lack of progress in implementing disposal concepts that are judged by many to be very safe and his unfaltering drive to improve this situation.

To provoke debate, the book is very deliberately written in a polarising, black and white style that can easily be labelled as "politically incorrect" - a characterisation that Piero will probably agree with and be amused by. Criticism is directed equally at "loud-mouthed and incompetent anti-nuclear environmentalists", the "nuclear Byzantium" of the international nuclear establishment, the "intellectual narcissism" of those nuclear experts that dare to admit the importance of societal issues, and the tendency of politicians to "indefinite procrastination". These are not words chosen to avoid open confrontation of opposing views.

His own sympathies are clearly laid out. He is no supporter of the Winston Churchill view that "scientists should be on tap and not on top". He firmly believes that experts alone - or at most with the input only of directly affected local people - should be able to determine repository concepts and sites. He firmly asserts views that are easily challenged, for example that "ordinary people are willing to .... respect, not distrust, technical experts" or that it is "completely false" that scientists in the past have worked in a non-transparent manner. He also ingeniously manages to use both the waste disposal success stories (as in the Scandinavian countries) and the project failures (as in Germany and Canada) as evidence that attempts to widen public participation are counter- productive.

This book then, is definitely not aimed at providing a rounded view of the world situation in waste management. On the contrary it will provide both supporters and opponents with ample opportunity to select points strengthening their case. Nevertheless, it is a text that can be very usefully read by both extreme groups, as well as by the (largely forgotten) uncommitted group in the middle. The book provokes one to think on the issues presented - and these issues involve key environmental choices facing society today. Like many other of his long-time colleagues in the field, I share Piero's frustration and I admire very much his continued dedication and commitment to moving things forward more quickly. With truly Italian passion, he wants to do all he can to ensure that what he believes to be right in waste management will actually be done in Italy, in Europe, in the world. I also share most of his views on the end objectives - namely properly designed, sited and constructed geological repositories that will ensure the safety of present and future generations.

What I do not share with Piero - as we have discussed at many a dinner table - are his views on why radioactive waste disposal has been such a controversial and lengthy process or on how the chances of more widespread and earlier success may be improved. Examples of arguments with which I cannot concur are: experts know best; wide societal interactions are counterproductive; siting should be purely technical; subjective perception of risks is of little or no importance. On the other hand, there are many points that deserve to be emphasised. That the nuclear waste debate is heavily influenced by the broader political agendas of participants, that huge efforts have been invested to ensure that repositories will be safe, that more should be made of risk comparison to other technologies, and that societal resources could be better invested in reducing risks elsewhere, are examples.

The fact that I have, despite diverging views on some issues, been invited by Piero to write this foreword demonstrates clearly his willingness, even eagerness to encourage open debate on all relevant issues. More important than the differences are the commonalities. For me personally, the most important message that Piero brings in his book is that scientists and engineers involved in waste management have been too silent for too long. Many have not spoken out in defence of the technologies that they have helped develop; many have restricted their views to effectively closed scientific committees and conferences. Piero's passionate arguments persuade me, not that less weight should be given to societal aspects, but rather that the scientists should become more directly engaged in the wider societal debate on the application of their technologies.

In practice, this is the course actually being followed by Piero in his unceasing efforts to progress waste management - a fact clearly demonstrated by this instructive and entertaining book.

Charles McCombie

Contents

1. The problem

Let's imagine a scientific meeting, or a committee of experts, convened to discuss and decide on the design and the construction techniques for a large bridge, planned to link an island with the mainland. The bridge – an important element – must be erected, because there is no alternative way of definitively solving the problem of communicating between the two sides. In other words, what is under discussion isn't whether to build the bridge or not – just how to proceed.

The very heart of the technical debate, for which distinguished specialists from several disciplines are gathered, ranging from construction theory to geology, is about selecting the best and safest solution for the bridge. Two options are being discussed, both dealing with the problems of statics, geotechnics and hydrodynamics – all of which are routinely involved in building bridges of this kind – and in this particular case concerning the bay and the emplacement of the piers.

These, as we can see, are purely technical matters, about which only scientists and engineers are capable of debating and deciding. At this point of the discussion, which, by the way, has been going on for a long time, all the required data are available: calculations and models have been made, the behaviour of the rocks and the maritime currents have been studied, and the seismic risk assessed. As matters stand, everyone is convinced that a sound and totally safe solution is on the table, and that the bridge would also be scenic and rather spectacular. There is nothing left to do but decide.

Surprisingly, when the moment comes to finalize the matter, the majority of the convened experts agree on one further point: in order to select the best technical solution it is essential to consult the local people.

Why? Because for the final decision it's a good thing that there is public consensus. Deep inside, some are convinced that the people have no knowledge about construction theory, but they remain silent, because are afraid of appearing antidemocratic, and because they begin to feel discouraged.

The meeting concludes with a final document in which, to stress the concept, pompous wording on democracy and participation is abundant, and where the engineers and scientists carefully avoid mentioning that the bridge could be constructed, that it would be safe and would resolve many problems. Furthermore, in order to give the people time to understand and digest the decision, they declare that several technical points still require clarification before a safe bridge can be constructed.

What we have imagined above regarding the bridge may appear odd, but it resembles quite closely what is happening nowadays with regard to nuclear waste.

These wastes have the peculiarity, which is now well known, even by the public at large, of remaining radioactive, and consequently potentially harmful, for a very long period of time after production. Consequently, to get rid of them it is not sufficient just to phase out nuclear energy. Even if, from some people's viewpoint they should have never been produced, they do exist, and unfortunately they will not simply disappear. Moreover, they have been generated to provide energy, for scientific research and for medical diagnostics and therapies, which means that they are the by-products of activities carried out to provide widespread benefits. Therefore, nuclear waste has to be disposed of safely. Doing this is not just an option, nor a simple precaution: it is a technical and environmental necessity. In other words it is a moral duty that cannot be overridden, and it is a fact that nobody questions .

If we compare this situation to our bridge debate, what is under discussion is not *if*. Besides, unlike the case of the bridge, even *how* is not under discussion, as technical solutions are available to dispose of the waste, which are under no circumstances disputed.

All the experts in the field, in this case too coming from many varied disciplines, are aware that the systems being applied or planned for nuclear waste disposal are absolutely safe. They are also aware that there is no waste produced by human activity today and probably not in the future, which can be disposed of so safely and reliably as nuclear waste. They are equally conscious that, to attain this objective, science and technology have made all feasible and reasonable efforts, including some, as we will see, that will possibly prove unnecessary. They are efforts that have never even been attempted for other industrial activities, including some having a far more dangerous impact on human health and on the environment.

Nevertheless, when the moment came to declare not only how to dispose of the waste, but also *where* (which also has to be technically correct and scientifically justified), our experts have been obliged to interact with the political decision makers, and wait for their decision. As this has not occurred, and probably will not in the immediate future, because politicians do not like making this kind of decision, which definitely does not contribute to their popularity, engineers and scientists have stepped backwards, losing their self-confidence and quite often their power of speech.

Without saying it openly, they have practically surrendered to the will of politicians, more or less consciously supporting the needs of politics.

Not unlike the imaginary experts convened to decide how to construct the bridge, these nuclear engineers and scientists have ended up convincing themselves that a decision on waste disposal – a decision which, under the circumstances, should only be one regarding *where* – can only be made when the people's consensus is obtained. Moreover, as if this were not enough, some, as we will see, have gone even further, and claim that the people should also decide *how* to dispose of nuclear waste.

Radioactive waste management experts had always been fully aware that selecting a site for a disposal system would have required careful and patient discussions with the local population, mainly based on scientific information, directed to acquire a certain level of consensus, or at least to minimize dissent. They also knew that any final decision would require a political consensus, achieved and verified by the usual instruments of democracy. But now, with the acquiescence and sometimes the contribution of engineers and managers, a new principle has arisen, apparently noble and truly democratic, which declares that the consensus *of society* is needed to dispose of nuclear waste. This being a concept difficult to define operationally, it has quite frequently ended up meaning the consensus of everyone, which as we know, is a highly difficult goal, especially in a democracy. Whilst waiting to achieve this, politicians can, of course, postpone the uncomfortable decisions that they so dislike making.

What has happened in nuclear waste management, beyond the technical achievements, to justify the present need of a societal consensus for implementing disposal, which is so convenient for those who are reluctant to make a decision?

It is a fact that, for about a decade, the problem of nuclear waste disposal is no longer addressed using rational and scientific criteria, as is the case for other industrial and environmental problems. On the contrary, it has become a matter lying somewhere between social psychology and cosmology, wherein arguments are admitted that elsewhere would undoubtedly be rejected as byzantine.

If this occur today for any type of repository for radioactive waste, it is geological disposal which is specially dominated by this attitude. As we will see later, this kind of disposal is planned principally for the so-called long-lived radioactive wastes, requiring such an extensive period of isolation that it can only be provided by certain deep geological formations.

The question first arose as to whether it is ethically correct to solve the problem of long-lived waste today, or to leave it for future generations. Then, others argued that it is not established whether safe isolation should be provided for one million years, or simply for hundreds or tens of thousands of years – ending up with the present day consideration about the convenience of allowing our descendants to retrieve the disposed waste in a remote and indefinite future.

All of this, it is worth mentioning, has not been devised by politicians but by technical experts, assisted by some sociologists whose attitude towards nuclear energy was not impartial.

What is the origin of this quiet and slow deviation, which could steer nuclear waste disposal into a *cul-de-sac* and even seriously jeopardize the future of nuclear energy, of which engineers and scientists have more or less consciously been the pilots?

***

In the last fifteen years, right when studies on nuclear waste disposal were approaching maturity, scientists and engineers had to face the new and unexpected situation that arose after the Chernobyl accident, when the antinuclear environmentalists made their greatest effort definitively to demonize nuclear energy.

However, the Chernobyl accident, which occurred in 1986, was the turning point not only of the antinuclear struggle, but also of the way that nuclear experts perceive their role and activities. For them, a kind of penitence has begun, more self-determined than imposed, as if they should publicly expiate the sin of having shown complete confidence in the safety of nuclear energy and having believed that such an accident could never occurr.

Nevertheless, even though it was understandable and appropriate that such an event generated uncertainty and doubts in the nuclear community, it soon became evident that what had happened to the Soviet reactor should by no means discredit nuclear technology itself, nor cause a crisis of conscience. Then the question is, what has led nuclear experts to this penitence, a kind of enduring Lent, in which they are no longer allowed to speak openly in favour of nuclear energy, and whereby they feel obliged to demonstrate doubts about their technical achievements? As if they should pay in some way to regain credibility and be politically acceptable.

In the democratic countries, the situation following Chernobyl has not only been dominated by emotional factors, amplified by the boisterous behaviour of the antinuclear environmentalists. Politicians have also played their role, both at national levels, and, as we will see later, in some international organizations. It is a fact that the root causes of the Chernobyl accident, which are far more *political* than technical, and of a typically Soviet political nature, were kept from the public at large.

It is, by the way, rather paradoxical that the need for transparency in anything dealing with nuclear activity, invoked today by technical experts and, equally, by politicians, was not put into practice for this particular case, which was, in nuclear history, the father of all events.

Yet the general public, the same public to which some want to give the authority to say where, or even how, nuclear waste should be disposed of, is far from being aware of the true causes and consequences of that accident. In the years following the Chernobyl accident, nuclear technology has been put on trial, with, on one side, as prosecutors, loudmouthed and incompetent antinuclear environmentalists, who are permitted to pronounce stupidities, and on the other side, as defence attorneys, technical experts and nuclear managers who were resigned to respond adequately and even to produce concrete evidence to obtain acquittal.

For years, the only circles where it was conceivable to remain unbiased towards nuclear energy were in the schools of nuclear engineering.

The most important consequences of failing to defend nuclear energy (which prevails all over Europe, but has led to a definitive outcome in Italy alone), have been on waste disposal. This may appear rather odd, as, in the nuclear fuel cycle, this is the activity having the lowest associated risk, by no means comparable with large plant operations, like power or reprocessing plants.

As a matter of fact, this has a twofold explanation. First of all, waste disposal requires evaluations extending, as we will see later in some detail, over very long periods of time. These evaluations inevitably involve non-technical issues, relating, for example, to the evolution and destiny of mankind, on which sociologists and philosophers, not being obliged to be scientifically correct nor consistent, may pontificate. This situation allows nuclear energy opponents to spread fear and worry, especially if technical experts remain silent .

Further, failure to provide a disposal system simply means that the storage time of waste must be extended by making additional space available, which does not significantly affect the economy of nuclear energy. This is the case for nuclear countries and, to a greater extent, for those having phased-out nuclear energy generation.

This explains why politicians do not feel any urgency to resolve the problem and to face the opponents of disposal. Instead, they are determined to prolong and defer the process of making the necessary, but disputed, decisions. In so doing they are encouraging, sometimes even claiming, the necessity of societal involvement and general consensus building, which may become an endless process. Deferring a solution is also serving the objectives of certain environmental parties, in governmental power in some European countries, whose political programme is mainly based on opposition to nuclear energy. It is essential for these groups to demonstrate that this energy is unable to resolve the fundamental problem of the wastes. Germany is a clear example in this sense, as we shall see later.

***

The weak defence of nuclear energy after the Chernobyl catastrophe, which we know had a massive impact on the perception of nuclear civilian technology, is not without explanation and justification. Nevertheless, 17 years after the event that left such a profound scar on the history of nuclear technology, we have to recognize that the fact that scientists and engineers are surrendering to the demands of politicians, is making solution of the waste disposal issue increasingly difficult. This is why it is now the moment to give back the priority to science and technology, because these are uniquely involved in safety and environmental protection.

A worldwide problem in nuclear waste management is the siting of the disposal systems. It is commonplace to say that decisions about site selection unquestionably lie with the politicians. It is, however, our belief that although we give them this responsibility, they will never solve the problem. Politicians have to face elections every few years, and if what they have to decide is disliked by the electorate, they will always prefer to postpone such decisions, possibly saying that the people must be allowed further participation and that greater consensus has to be achieved before going ahead.

Nobody disputes the principle that locating a repository for nuclear waste has to be implemented without conflict, and, as far as is feasible, with the public's involvement, or that any final decision requires approval from the usual institutions of representative democracy. Nevertheless, people have to be informed that solving the problem of nuclear waste is not an option dependent on the democratic interplay of political parties, but an unavoidable civil obligation. It is the duty of politicians to honestly explain to people this simple truth, using the mechanisms that they judge to be most appropriate and effective, but avoiding opportunism and demagogy. The decision on *how* to provide a solution for waste disposal, the criteria to be applied for and, finally, *where* to locate the repository, should more appropriately be left to technical experts, who are accustomed to use the tools provided by science and technology, and not by politics.

It is not proved that ordinary citizens are more confident in politicians than in scientists and engineers. We are firmly convinced of the contrary, otherwise we should admit that either information has been manipulated by politicians' propaganda, or that technical experts have lost credibility because of their conformism.

It is, after all, in the interest of the politicians that decisions regarding a nuclear waste disposal site, undoubtedly doomed to create opposition and dissatisfaction, have to come from the technical experts. Even when linked to some political or economical circles, they usually maintain sufficient independence of judgment. Decisions made by technical experts, on purely technical grounds, usually have weaker political repercussions and consequently less significant impact on the destiny of politicians.

In order to contribute to restoring technical experts to their role (which in some ways is also a political one, but in a more general sense), it is useful to recall the main events and try to explain why the problem of nuclear waste disposal, which technically speaking is not hard to solve, has become the focus of the difficulties encountered by nuclear energy. We will also try to explain why nuclear waste disposal has become the opportunity for an ecological struggle dominated by irrationality and misinformation, where the needs of politicians appear to have prevailed over those of science, and even of environmental protection, with a consequent huge waste of resources.

2. A special energy and its enemies

The detonation of a charge of explosive, which occurs when firing a cannon shot, exploding a landmine, or in an aerial bombardment, is due to a chemical reaction (named oxidation) that conceptually and energetically is the same as the one occurring when heat is produced by burning kerosene in a boiler, coal in a furnace, or a log in a fireplace. What changes in the two cases is only a thermodynamic parameter, known as reaction rate.

However, nobody has ever considered associating the two phenomena, explosion and heat generation, and bringing chemical combustion to trial, questioning its use, or even setting off on a crusade against it, as one of the demons of technical and scientific progress. Even less has anyone considered placing the problem of the uses and effects of chemical combustion at the centre of the programme of a political party.

All this has happened with nuclear energy and continues to happen. The production of thermal energy in the core of a nuclear reactor and in the explosion of an atomic bomb takes place by means of the same basic reaction (which in this case is not a chemical but a nuclear one – because it involves not the electrons but the nucleus of the elements). But, as with gas burning in a stove and an explosive charge blowing up a building, the relationship between the two processes ends with the reaction mechanism, while the effects of the reaction are greatly changed and focused towards the desired result by means of technology.

The ambivalence between destructive and constructive uses is by no means a prerogative of nuclear energy, neither is it more conspicuous or evident than in the case of energy from chemical combustion. On the contrary, the scientific progress of the last 150 years in the controls and materials used in chemical oxidation processes has made possible the development of both advanced energy machines and increasingly powerful weapons – as a result of which the victims of wars in the twentieth century exceed those of previous centuries by many orders of magnitude.

On the other hand, this is not a novel issue in the history of mankind; man has always tried to twist any available form of energy into non-peaceful uses. The words of Seneca, in 62 A.D., in a famous excerpt from the *Naturales Questiones,* concerning the winds, may have relevance for all energy sources: "....they make the air breathable, bring rain and clear weather, transport seed and move sails over the seas furthering commerce and relations", and he added "immense benefit, which human folly has turned to ruin: it was not for this purpose that God, universal providence, had provided the winds, so as men may fill the ships with armed men and hunt down their enemies on the ocean and beyond the ocean.......".

The reasons why it is only in the case of nuclear energy that perception of this ambivalence has become widespread, and indeed has acquired a political and social significance, are only partly due to the terrifically destructive circumstances in which this form of energy first appeared to the world. After the Second World War ended, with pictures of Hiroshima still on the pages of newspapers and magazines, the race towards peaceful uses of nuclear energy started almost immediately, without any moral reservations or crises of conscience (the crisis of scientists such as Oppenheimer was not related to the peaceful exploitation of nuclear energy). Instead, the development of nuclear reactors and materials became one of the advanced areas of technological research in the major countries and, for more than twenty years remained, in the east as well as in the west, one of the most appreciated and unchallenged technical and scientific activities. The term nuclear, moreover, acquired a more general significance and in some cases was used to characterize both the epoch and its society. If we go back to look at some social inquiries of the 60s and 70s, we find expressions such as "nuclear era" or "nuclear society" to denote the historical and the social climate of the time, with its problems and interests. Twenty years after Hiroshima then, towards the middle 60s, the troubling world view, that peace was based only on a balanced atomic arsenal, did not prevent the most optimistic estimates regarding the commercial and industrial use of nuclear energy. Besides the well known projections on production of nuclear electricity, both scientific conferences and the pages of the popular press envisaged that, by the end of the century, automobiles would be powered by plutonium batteries.

Considering that, nowadays, that unfortunate element plutonium only evokes among the public visions of more or less dirty atomic bombs, or at best of radioactive pollution, one can measure the distance that the perception of nuclear energy has covered, in a period of time during which the original sin of Hiroshima, if it can be called so, could not evidently have played any role. Those who still evoke the inheritance of the bomb in analysing the problems of social acceptance of nuclear energy, particularly in relation to the siting of repositories for radioactive wastes, are far from the true origins of the phenomenon. These are to be found not in the connections between military and peaceful uses of nuclear technology, but rather in processes belonging to the genesis and history of cultural trends.

***

Towards the middle of the seventies, in the prosperous nations of the western world, militant environmentalism appears. Rather than being part of the fall-out from the youth movements of previous years, the defence of the environment, based not on practical but on ideological grounds, most likely arises from the failure of the communist myth which was becoming apparent in those years. Indeed, environmentalism, even though originating from needs and aspirations that have nothing in common with the class struggle (in fact, it used to be a prerogative of social and cultural elites), immediately acquired followers among the left and soon became a creature "of the left", which identified it as a potential battle field against the bourgeois society of the western world. Enlightened environmentalists disappeared rapidly from the scene, particularly in the USA, the country where the movement originated. Here, environmentalism inherited and merged with the *campus* culture, which had been the forge of youth opposition to bourgeois society and which, in many respects, is the basis of the European left's culture, much more so than Marx.

Having an ideological and opposition character, the rising environmental movement needed to grow a strong guiding idea. This could not be the struggle against classic pollution, from automobiles, urban heating or industrial chimneys. On the one hand, this environmental problem was too close to the common people and to their needs and aspirations, on the other it was clearly linked to the industrial power, which, besides oil and automotive world, controlled the means of communication, and whose enmity would have deprived the movement of the visibility in the media that they crucially needed. These environmentalists are, in fact, not revolutionaries, being neither similar to the improvident romantic strugglers of the past, nor yet comparable to the professionals that led more or less just and successful revolutions. Legitimate sons of the affluent society, they have absorbed, together with the life style, the cunning of the system. They are indeed well acquainted with the importance of communication. It is not by chance that Hollywood, as we will see, will not fail to give them support.

The strong guiding idea of the environmental movement was the hostility against nuclear energy. On closer inspection, all the ideological requisites were there. The obvious spectre of the bomb, especially when coupled with the ambiguous connections of peaceful and military uses of nuclear energy at a period of history dominated by the cold war, caused professional pacifists and supporters of nuclear disarmament to join forces with these environmentalists. The USA led the technological development and also were the main commercial supplier of nuclear plants and equipment, through two or three large commercial companies. This was not only disliked by the traditionally anti-American European left, but was also identified, both inside and outside the USA, with so-called *corporate multinational imperialism*. In order to guarantee safety and minimise risks, the users of nuclear energy proposed strict organizational and operational criteria, together with strict compliance with regulations; all this could pass, in a permissive society, for an *authoritarian governance* of the system.

There was another primary factor. The 70s were a time when there was open and unrestrained discussion at every scientific conference (and there were many of them) not only on the progress of nuclear technology but, above all, on the development under way in materials and processes aimed at improving both the safety and the economics of the system. On these conference papers, badly read and interpreted, the earliest nuclear opponents based part of their know-how and from these they drew part of their polemical weaponry. If they read, in a paper, that a more advanced process for conditioning radioactive waste was under development, they cried out that the waste problem was not solved. If they read that in a nuclear plant there had been a certain malfunction (a typical issue discussed in any respectable conference, where operating experience is usually reported), they declared that a nuclear incident had taken place, and, furthermore, had been kept secret. If they found it written that nuclear fuel could be upgraded and could produce, say, more plutonium, they hinted that it could be put to improper use.

(Many nuclear experts, repentant or obliged to repent in recent years, have beaten their breasts in papers and scientific committees, accusing themselves, as self-critics did during the Inquisition or the Stalinist trials, of having in the past worked in a non transparent way and of having practised a kind of esoterism. This is completely false: scientists and experts in the field of peaceful nuclear energy, even when representing commercial interests, have always reported the state of the technology without reticence during the scientific debates that took place at the time of the successful development of nuclear energy – at least in the democratic western world, as will be shown in the next chapter. Of course, they did this at appropriate venues, which, while not well attended by the general public, were free and open to all, and in no way restricted.)

Thus, the rising environmental movement identified the opposition to nuclear energy as its preferential battlefield. We will see later how the Chernobyl accident will dramatically speed up a process that was already underway, offering ideological environmentalism an unhoped-for weapon and placing an army of technical experts in extreme difficulty. However, since the beginning, a decisive factor in favour of the nuclear opponents must be noted, which confirms that they had chosen the right battleground: the benevolent attention that the media give them.

Even without indulging in the art of exploring the hidden causes of events, one cannot fail to observe that, in the years of rising environmentalism, the *actual* harm to the environment, due to the chemical industry and to fossil fuels, has not been highlighted in the media to anything like the same extent as the *potential* harm due to nuclear energy. Whether, thereby, militant environmentalism has been encouraged in its antinuclear crusade in order to keep it away from battles that would have disturbed consolidated industrial powers, remains a good question. In whatever case, it is a fact that environmentalists have only recently, having reinforced themselves politically, and with nuclear energy being in retreat almost everywhere, devoted themselves to conventional pollution problems. But they have done so at a time when the adoption of less polluting production processes has become commercially attractive. Anyhow, professional environmentalists have never dealt with conventional pollution, not even the most diffuse and deadly, with the same virulence that they have reserved for nuclear energy.

There was another factor in favour of the antinuclear crusade, on which the environmental movement founded its programme. The proximity between peaceful and military uses, which we have seen is a potential prerogative of almost all energy sources. In the nuclear case, this was bound not only to have an impact on the militant pacifists but also to influence large layers of western society, in a period of the 70s and 80s when international scenarios were dominated by problems connected with the balance between the superpowers' nuclear arsenals. An example of this is the political debate about the installation of nuclear missiles, which ended up by involving not only contesting minorities but also mass movements (particularly animated was the discussion in Italy). Besides, the *maitres à penser* of the dominating culture, mostly siding with the left in their religious (catholic as well protestant) and lay components, did not contribute to endorsing the *realpolitik* concept, but preferred to talk of the balance of nuclear terror. In this situation, the same intellectual *milieu*, willingly but seemingly unintentionally putting together commercial and military issues, liked to present the rich and affluent western world as the one giving the bad example.

These boundary conditions, as we say in mathematics, contributed to create in the public at large, and not only in its leftist side, if not a hostile climate, at least a nervousness about everything connected with the use of nuclear energy. Meanwhile, absolute silence covered the advantages of that energy.

What has happened in Sweden is emblematic. The country of neutralism, always in the foreground of pacifism (a stance by the way aided by geographical position), was the first in the West to officially oppose nuclear energy. Following a referendum held in 1980, it decided to shut-down all nuclear power plants by 2010.

The case of Sweden shows how the perception of nuclear energy is influenced by social factors and cultural trends. The approaching date established by the referendum is causing great political and industrial embarrassment, because the power stations are working perfectly, their safety is absolute, and moreover the opinion polls show that diffidence over nuclear energy, once the pressure that generated it is vanished, is disappearing. Even the memory of Chernobyl is fading away, as the psychological climate is changing. But we will return later to the case of Sweden.

In the western countries, press and television, especially when their orientation was liberal (as will be shown later, even the so called committed cinema has given a hand), contributed decisively in forming this vigilant and suspicious attitude towards nuclear energy. This was not only done indirectly, by almost completely disregarding the environmental issues connected to conventional industrial activities mentioned above, but also by almost always handling nuclear energy issues as a "scoop" in which there was no intention to inform or to promote a rational appraisal.

At the beginning of the 80s, on the other hand, a style and culture prevailed in communication, including advertising not less than political information, which addressed the public's emotional attitudes more than the rational ones. Colour, editing techniques, written and spoken headlines, use of photography, sound tracks, everything is directed to strike the receiver of the message rather than to inform him (this is why some speak of impressionism being applied to information). The most immediate victim was bound to be the understanding of the issues.

The antinuclear ecology that grew during those years, generating scares and spreading illusions (such as those concerning alternative energies, not by accident called "soft"), had all the requirements for taking advantage of this prevailing information philosophy. In fact, the media has since shown not only sympathy and understanding for the first battles of the nuclear opponents, but also, aside from a possible submission to the strong industrial powers responsible for conventional pollution, a complete lack of interest in having a neutral discussion about nuclear energy, leaving aside those alarms and doubts that they considered essential to raise public interest.

Since then, it has effectively been impossible to discuss the advantages and dangers of nuclear energy with an unbiased mind. This situation has not been helped by the underlying *political polarization* of energy, which dates more or less from that period, and took place particularly in Italy, Germany, Holland and the Scandinavian countries, according to which – here we simplify - alternative energies belong to the Left (in other words to the progress side) and nuclear energy to the Right. Alternative energies are sweet, fair and almost homely (like Sundays with no autos), while nuclear is steely, heavy and authoritarian. It is interesting that fossil fuel, the really essential source of energy, found no place in this polarization. Even though responsible for vast environmental damages it could not bear the luxury of ideology, as with all things that one cannot do without.

If the above mentioned polarization mainly belonged to the ideological environmentalists, it also found a home in the traditional left, as shown by these opinions of well-known experts from the Italian Communist Party, (S. Bologna, G. Cesareo, M. Pinchera) on the occasion of the hot debate on energy which took place in Italy at the end of the 70s.

> The nuclear choice appears as the final achievement of a whole historical phase of the capitalistic countries' energy policy: it is indeed consequent to the choices made before it, and aligned with the logic governing the development strategy in these countries. This choice is organic to the capitalistic mode of production, which it contributes to preserve, on one hand revolutionizing its technologies, on the other enhancing its distortions. For this rea-

> son the nuclear choice must not be considered by itself, but within the context of national and international class relationship: it must be considered not an economical and technical choice but also mainly a political one. A choice dictated by imperialism.
>
> [...] Actually, the one realistic way to aim at the transition appears to be the concentration of research investments for the prompt development of alternative energy sources in a democratic and decentralized manner [...] in order to intervene on the capitalistic mode of production and on the social organization.

The political polarization meant that, for years, a great part of the so-called progressive press gave credit, more or less openly, to the idea that there was no clear border between the civil and military uses of nuclear energy (especially when the USA was involved), and to the equally misleading idea that alternative energies were economically and industrially viable. It is only after Chernobyl, though, that the troops of the left, especially in Italy and Germany, deploy themselves decidedly and unanimously against nuclear energy.

Because of the new trend of information on the one hand, and political polarization of energy on the other, nuclear energy has therefore never benefited from neutral information, which instead has been shown for almost all the other areas of science and technology. We say almost, because, in the meantime, militant environmentalists have identified two other domains where they can carry on with their ideological ecology and rely on the media already showing adequate understanding of their position: biotechnologies and weak electromagnetic fields.

In the case of nuclear energy, it will probably no longer be possible to convince public opinion of the distance separating the reaction that takes place in the bomb from that running a nuclear power plant, as is the case for a mine exploding and a gas fuelled power plant. Likewise, it is not easy to explain how and why in the case of nuclear energy, with respect to other human activities, the risk is not greater, but, for precise historical and sociological circumstances, the perception of the risk is different.

However, beyond its history and the circumstances of its appearance, besides the problems made by the ideological opposition of the environmentalists and by the methods of the media, nuclear energy continues to be very special. Compared to other forms of energy, it presents technical, economical and industrial aspects which are very specific, and which deserve to be discussed.

***

When natural gas, oil or coal is burnt in a conventional power plant, the fuel disappears almost completely, in the sense that it almost completely turns into smoke. In the case of coal and oil fuel, solid residues of combustion remain, similar to ashes. Their volume is negligible with respect to the initial volume of the fuel, but their content is extremely toxic.

Instead, the fuel feeding a nuclear plant is an industrial artefact called a fuel element, comprising a metal structure containing the energy releasing material (the actual uranium fuel), which produces heat inside the reactor (boiler) and is discharged without any physical change after having supplied a certain amount of energy. This amount, an important detail, is just a small part of the energy potential of the fissile material of the initial charge. When the fuel element is unloaded from the core, it still contains a large quantity of material capable of producing energy. We will see later that this quantity in some cases is larger than the initial one. The role played by nuclear fuel, evidently very different from that of fossil fuel, has originated a specific sector of nuclear technology, which has been called the *fuel cycle*, with a corresponding specialized industrial and commercial sector, parallel and complementary to the reactor sector.

In the core of the reactor, part of the uranium undergoes a nuclear reaction, which produces, besides the thermal energy used to heat water (or gas), new chemical elements, unstable from a nuclear point of view and therefore highly radioactive (radioisotopes). These remain confined inside the high integrity rods of the metallic structure of the fuel element. The penetrating radiations associated with the new isotopes make the fuel element very radioactive. Therefore, from the moment of its discharge from the reactor, it has to be handled remotely, with special caution.

While for a gas or oil fuelled power plant the technological and economic life of the fuel ends with its combustion, for a nuclear station everything begins, so to speak, once the element has done its duty in the reactor and is removed. In the last twenty years, great attention has been paid to this phase of the fuel cycle (called the *back end*), which follows the fuel's residence inside the reactor, on account of its technical, economic and also, as will be seen, political aspects.

In the years of the development of civil nuclear energy, during the 60s and 70s, when commercial nuclear reactors achieved full maturity, most of the technological effort was on the activity directed to produce the fuel to be loaded in the reactor (the *head-end* of the fuel cycle). In order to obtain an acceptable energy output and avoid any release during the nuclear reactions, and thence to guarantee safety, it was necessary to identify chemical and physical forms of both uranium fuel and structural material, that allowed a sufficiently long residence of the fuel element in the reactor before being discharged. With the 80s, the parallel development of fuel element fabrication techniques and the design of more reliable reactor systems, resulted in western countries and Japan identifying a class of commercial reactors, the *light water reactors*, that have been widely employed (only in United Kingdom and Canada have different types of reactor been adopted). Following the achievement of a mature and standardized fuel-reactor system (with a high level of reliability), attention was transferred to the other sector of the fuel cycle, dealing with the management of the fuel unloaded from the reactor after having produced energy, technically called *spent fuel*.

In spent fuel, we can find everything that makes nuclear energy special, and which today is at the root of most of the problems connected with this kind of energy. In particular, it is the spent fuel management, i.e. the operations it undergoes when removed from the reactor, that definitively characterizes both radioactive waste production and the entire energy cycle.

The radioisotopes produced during irradiation in the reactor, remaining inside the discharged fuel element, are of two main types: the so called fission products, and plutonium, with its possible derivates. If chemically separated, the first become radioactive waste and the second an element capable of producing energy having the same characteristics as fissile uranium (the so-called uranium 235 isotope, 235 being its atomic weight).

Before discussing spent fuel and radioactive waste further, we have to pause for a moment to say something about plutonium, in order to clarify the technical aspects of the production of this element, which is too often brought up and discussed in a completely unrealistic context.

Plutonium is an artificial element, not occurring in nature, obtained by irradiating uranium in a reactor. Like fissile uranium, it can be used for energy production but also, in certain conditions, for military purposes. However, there is an important difference between the two uses: the plutonium produced in a commercial fuel element is not suitable for a bomb. Its isotopic composition cannot ensure an efficient explosion. Besides, it becomes increasingly unsuitable the more energy the commercial fuel has produced. The two requisites, energy production and military use, are thus in conflict.

It makes no sense to produce plutonium for military purposes by using commercial fuel and commercial reactors, and nobody has ever tried to do it. It may be possible of course, in a commercial power plant or in a research reactor, to introduce a special fuel element and irradiate it in a particular manner in order to produce plutonium, which if not of weapon grade, may be used to make a weapon of lower efficiency but still dangerous (a kind of dirty bomb). But even with this 'homemade' way of producing plutonium there are two obstacles that are hard to overcome. First, the separation of the plutonium produced during irradiation is carried out by a chemical and mechanical process – so-called nuclear fuel reprocessing - performed in plants that are the most complex in the fuel cycle, and which exist in only a few countries. These countries, on the other hand, possess nuclear weapons, and operate dedicated plants for the production of suitable fissile material, fully independent from the commercial ones. Second, an elaborate mechanism of international controls, under the United Nations through the International Atomic Energy Agency (IAEA), supervises production and transfer of fissile materials, including those for commercial use, in all countries that do not have nuclear weapons. These controls are accepted voluntarily by the countries that have signed the Non Proliferation Treaty (NPT), which imposes safeguards for fissile materials.

The last two countries to develop the atomic bomb – India and Pakistan - have not done so in a clandestine way. They officially and programmatically refused to sign the NPT, having equipped themselves, at considerable expense, with specific, probably small scale, plants for the separation of plutonium and for its further manipulation. In short, acquisition of plutonium is not within the reach of any ill-intentioned country, as has often been asserted in order to create distrust in nuclear energy.

To acquire fissile material for weapons, it would be a less forbidding route to enrich uranium by increasing artificially the percentage of uranium 235 in natural uranium. This is proved by the fact that, in Iraq, after the Gulf war of 1991, equipment for the production of enriched uranium was found about ready to be assembled; the equipment was obsolete but fit for use. Recently, a black market of enriched uranium from the countries of the former Soviet Union has been reported, but the trustworthiness of the information is uncertain, as it comes from a press release.

Wholly different and certainly unable to evoke war scenarios, it is the issue of the energy potential of plutonium. In the nuclear fuel that produces heat in a power plant, while part of the uranium is burnt, another part produces plutonium, a material with a great energy content, and better able than uranium to sustain a nuclear reaction and produce thermal energy. Part of the generated plutonium is directly consumed in the reactor by nuclear reaction (thus producing energy), while the rest is to be found in the fuel removed from the reactor. In particular circumstances, the quantity of plutonium formed in the spent fuel can be larger than the original fissile material. These circumstances occur in fast reactors, so named because the neutrons in the core are not slowed down as in thermal reactors. We thus have a so-called breeder cycle, because a fissile material is produced that is capable of keeping the cycle running without addition of fresh material. However, the fast reactor cycle requires that the initially charged fuel be plutonium-rich: this means that a given quantity of this material is needed for the fabrication of a fast reactor fuel. There is no other way to get it but recover the plutonium by reprocessing the spent fuel in which it has been generated during irradiation in the reactor core.

The evolution of nuclear technology and the actual history of commercial nuclear energy, during the first decades, at least up to the end of the 70s, have been dominated by the really appealing perspective of having an energy cycle that did not need raw materials, when at steady state conditions, but only technological capability.

It was essential in this perspective to set up industrial scale fast reactors, an endeavour much more complex than installing thermal reactors, on account of the engineering and safety problems involved. Towards the middle of the 60s, in all the major industrial countries a race took place to develop semi-industrial scale fast reactor prototypes. The wars in the Middle East, with their repercussions on the oil price, were a powerful accelerator for fast reactor development programmes, besides furthering the installation of the thermal ones.

Also essential for the breeder cycle was the industrial development of spent fuel reprocessing, without which plutonium could not be recovered and complete energy independence could not be attained. This gave rise to a vast technological and applied research effort on reprocessing, undertaken from the 60s in all countries using nuclear energy, in particular in those highly dependent on fossil fuel sources (Italy built two pilot plants).

The first country to shed its illusions about fast reactors was the USA. Already in the middle 70s it appeared evident, to anyone who wanted to see, that the advantages connected to the breeder cycle were quite uncertain. There was none on the economic side, since both the development of commercial fast reactors and the industrial reprocessing of the spent fuel encountered prohibitive and continuously rising costs, on account of the increasingly severe safety requirements established by the competent authorities. Therefore, the breeder cycle, though autonomous in theory and prospect, began to look economically unattractive if compared to thermal reactors. Besides, the possibility of energy autonomy was of diminishing interest, because there were no problems, not even in the long term, for the supply of uranium minerals. Their reserves, always more abundant and practically inexhaustible, were located in seemingly zero-risk countries, such as Canada and Australia. With thermal reactors, it therefore seemed possible, at a much lower cost and without great effort, to obtain the same results – diversification of energy sources and freedom from unreliable suppliers.

In 1976, the USA suspended commercial spent fuel reprocessing and, shortly after, in 1979, they cancelled the fast reactor programme that had been proceeding sluggishly for several years. In Europe, some found hidden motives behind these decisions: the intention of the USA to stop the production of plutonium on account of the dangers of proliferation was considered. As a matter of fact, during the Carter presidency this problem could have been taken into account, but the decision was maintained by the Reagan administration and thereafter. The American decision simply reflected the belief that fast reactors were not good business. The same conclusion was reached by the Europeans ten or fifteen years later. Some countries, like Italy, had to wait for Chernobyl to definitively cancel programmes that had swallowed up huge resources and had no future.

The case of fast reactors is quite similar to that of civilian supersonic aircraft. In this case too, the USA, after conceiving the project and setting up the first studies and prototypes, abandoned the initiative. Instead, it aimed at large subsonic carriers and, in this case too, now notoriously, it was found that the cancelled project was not good business, although others carried it on just to keep the flag flying.

In the USA the decision to stop the fast reactor programme caused commercial fuel reprocessing to be abandoned, and large and very expensive plants, already built, never started operation. It was acknowledged that this operation had a technical and economic sense only if aimed at recovering plutonium in order to reuse it. Without this, fuel discharged from the power plant had to be disposed of, in spite of its energy content, actually becoming a radioactive waste, the main industrial waste of the nuclear energy. This fuel cycle was called the *throw away* cycle. In Europe and Japan, fuel reprocessing continued, and continues today, even after fast reactors programmes have been officially dropped. Why?

To begin with, it is to be noted that in Europe, and in particular in the two most important nuclear countries – France and United Kingdom - the fast reactor perspective survived during the 80s and afterwards. Its survival was due to a kind of necessity: when the economic weakness of the fast reactor option became evident, the two countries had already built large reprocessing plants to satisfy European and Japanese demand, requiring colossal investments. Moreover, the potential customers had shared financially in the investments, in exchange for reprocessing services for the fuel coming from their national power plants. Fuel reprocessing, therefore, continued to remain in the official programmes of several countries simply in order to justify the economic resources that had been invested. Or, to put it a different way, having invested in reprocessing, these countries looked less pessimistically at the future of the fast reactors.

But above all, those countries have always tended to give credit to another philosophy regarding spent fuel reprocessing: the impact that this operation has on radioactive waste management. It was due to this philosophy (and the capability of reprocessing to separate plutonium) that nuclear fuel reprocessing became an industrial process having ideological opponents.

***

As we have seen, newly formed radioactive isotopes accumulate in nuclear fuel during its irradiation in the reactor. About 98% of all the radioactivity generated in the reactor is inside the metallic structure (fuel element) discharged from the reactor after having produced thermal energy. The remaining fraction is in the radioactive waste produced during the operation of the plant itself (mainly in the purification system of the hydraulic circuits that removes heat). For 98% of the waste coming from a nuclear reactor, the problem of managing radioactive waste is the problem of managing spent fuel.

It is evident that, even though the inventory of the radioactive material to be disposed of is the same, its management will be quite different if the fuel element containing it is disposed of as a whole, or if it is reprocessed, which means cutting it in pieces and processing it chemically. What, in the case of direct disposal, remains confined in the metallic structure of the fuel, with reprocessing is distributed among various streams of liquid and solid material coming from the plant. Volumes, weights and types of the final materials containing the radioactive substances (the actual wastes), are consequently very different in the two cases.

The principal and most complex problem regarding nuclear waste is that of the high activity, or long lived waste, containing a large concentration of radioisotopes whose decay time is in the order of thousands or tens of thousands of years. If reprocessing is not carried out, there is only one type of radioactive waste downstream of the reactor - the discharged fuel element itself, which contains all the activity. Therefore, it is routed for geological disposal, after a period of storage. In the case of reprocessing, the material discharged is divided by plant operations into high and low activity waste; the latter, notwithstanding its large volume, is easier to dispose of, because the disposal can take place in engineered near-surface structures, as will be shown later in some detail.

As for the volume, that of the high activity waste from reprocessing is lower by a factor of about four if compared with spent fuel. The waste consists of blocks of glass in steel containers, possessing high stability, chemical inertness and physical durability - even more than the unreprocessed spent fuel element.

It is actually this limited advantage in favour of the disposal of vitrified high activity waste with respect to the whole fuel element (smaller volume and greater durability), that was and is being emphasized by those stressing the convenience of reprocessing fuel, even when there is no need for recovering plutonium as a nuclear fuel. Like the reasons that delayed abandoning fast reactors, this argument was also dictated by the understandable need to defend and justify the strategic choices of the past and the consequent massive investments in building reprocessing plants.

If it is not necessary to recover material capable of producing energy, it is senseless to submit spent fuel, having *per se* a robust and long lasting structure, to complicated mechanical and chemical processes, to destroy and disperse it in various liquid and solid waste streams that must be reconverted into a stable and inert material, adequate for disposal. Paradoxically, it would not be convenient even if the two options, reprocessing and *throw away* cycle, were economically equivalent. Instead, the reprocessing option is much more expensive. Furthermore, reprocessing generates the problem of safely storing the plutonium recovered by reprocessing. If not re-used, plutonium is also a nuclear material that must somehow be disposed of, and this is by no means a simple matter.

The advantages of disposing of high activity waste are also more theoretical than real in case of reprocessing. The smaller volume to be disposed of and the greater stability of the material in which the waste is incorporated (the so-called *waste form*) do not have a significant impact on the design of the geological repository, because long term isolation is considered to be largely assured by the natural medium and not by man-made structures (this is also true when natural and man-made structures have to work together, like is supposed to be in the Swedish bedrock).

France, the country that has, with great determination, pursued nuclear fuel reprocessing and possesses plants of enormous capacity with the most up-to-date vitrification technology for high activity waste, continues to be the most active in looking for and defending the advantages of this industrial option. But it is again, as in the past, a flag flying battle, not unlike continuing the commercial flights of Concorde.

Nobody today in France or in the United Kingdom thinks seriously of continuing fuel reprocessing into the middle term future and all the other European countries have given it up, or will give it up as soon as they are free from the contractual obligations with the plants in these two countries. Spent fuel reprocessing, as an industrial practice, is dead, but it has not been buried yet. As long as nuclear electricity production continues, using current technology, there will be a need for direct disposal of spent fuel.

For at least 30 years, spent fuel reprocessing has been a peculiarity of nuclear energy, more than any other of its processes. A generation of engineers and chemists has dedicated itself, in all industrialized countries including Italy, to its scientific and technical problems, which are among the most stimulating in all nuclear technology. When, at the end of the 70s, professional environmentalists tried to find a more technical field on which to focus their noisy arguments against nuclear energy, reprocessing, which in those years was reaching industrial maturity, captured their attention. And they made their first bugbear of it.

If, as we simplified above, for ideological ecology, nuclear energy became the energy of the Right, reprocessing was to the right of the Right, while direct disposal of spent fuel became, so to speak, the left of the right. How can this be explained?

Considering that we are talking of ideological environmentalists, the reasons are obviously not to be found in the technical and economic logic of the fuel cycle (which actually has put an end to reprocessing), nor in the real environmental aspects of the two radioactive waste management options we have discussed above. At that time, notwithstanding the doubts existing among experts and the decisions being envisaged in the USA, the breeder cycle and fast reactors still publicly shared a vast credit. We would say that upon them converged the most futuristic, almost messianic, expectations regarding nuclear energy and energy independence. To those who were unable to investigate the problem closer, reprocessing appeared to be the natural target in order to hit a sensitive point of this promising energy cycle. In view of this, direct disposal of spent fuel, just because it made impossible to recycle fissile material, became a good thing.

The obvious polemical tools used to attack the reprocessing were, on one side, plutonium recovery and its possible diversion to non-peaceful purposes, and on the other, the large quantities of radioactive waste produced and the lack of satisfactory criteria for their long term management.

Polemical puppets, as we have tried to explain. But, before Three Mile Island and, moreover, before Chernobyl, nuclear energy opponents had very little else available to attack.

3. Before and after Chernobyl

The ideological environmentalism founded on the opposition to nuclear energy arose and spread throughout the western world during the seventies, in a period when nuclear technology, both for reactors and the fuel cycle, was approaching full maturity. It arose and developed, as we discussed in the previous chapter, in the wake of and as an additional form of protest against capitalistic society, and not because there was social concern about civilian uses of nuclear energy, a concern which did not exist at that time. Events such as Three Mile Island and Chernobyl came in fact later. What are, then, the topics on the agenda and the heroes in this pioneering stage of the antinuclear struggle?

In an American film from 1979, an almost unknown Michael Douglas played the role of a cameraman who can be considered the prototypical 'antinuclear'. The film was "The China Syndrome", and, in a sense, it represents a real *manifesto* of the antinuclearism. In this film we can quite easily identify the features and original positions of the opponents of nuclear energy, which, as we will see, have very little if anything to do with environmental protection, but very much to do with criticism of certain faults and vices of capitalism.

One point should be mentioned before going further: when the film began showing in cinemas, the Three Mile Island accident had not yet occurred, so the story could not have been inspired by that event. The accident occurred three weeks after the *première*, and it was only from this moment onwards that the film became a real box-office success. The China Syndrome is clearly a politically oriented film, as was demonstrated by engaging a star like Jane Fonda, who was a notorious activist committed to "democratic" battles.

Jane Fonda plays the role of a TV reporter who, supported by Michael Douglas' cameraman, is about to make a report on a nuclear power plant. Having nothing against nuclear energy, she is favourably impressed by the complexity of the plant and its well managed organization. The cameraman, on the contrary, is not. He is convinced that he is entering the Devil's den. He is rebellious and excitable and, once in the nuclear plant, becomes suspicious of everything, from the controls at the checkpoint to the direction to wear a protective helmet. Extremely circumspect, he seems worried on hearing any unfamiliar sound or a loudspeaker announcement. We can imagine what he is thinking when, looking at the activity under way in the glass-walled control room, he realizes that something is wrong over there. The control room is sound-proof, but he notices an abnormality as warning lights flash on the control panel and he sees the worried looks on the faces of the operators. The person most affected appears to be the supervising engineer, played by a rather excited Jack Lemmon (as we can see, the politically correct Hollywood takes the field with a team of real champions!).

The visitors are escorted by the public relations manager, who seems to have an institutional duty of hiding things from them. As we would say nowadays, he is the personification of non-transparency. Even his face is unpleasant (apparently, in the nuclear business, it is possible to employ such a fellow in public relations!). This man also realizes that there is a problem in the control room, but he reassures the visitors, telling them that everything is *routine*, and explaining to the cameraman that, for commercial confidentiality, filming is not allowed in the control room. Doing his utmost to expedite the visit, he finally conducts the visitors to the exit, saying good-bye with a sigh of relief. However, something escaped his notice: the cameraman has kept his camera turned on during the time spent looking into the control room, and the dramatic moments inside have been fully recorded.

The truth is that, in the control room, they have been quite close to a really dramatic situation (and the audience is aware of this, unlike the two reporters who were outside). Because of a trivial malfunction of a water level meter (which seems rather incredible for a nuclear plant), the reactor's cooling circuit was being drained, a situation that could have made the temperature of the core uncontrollable. Fortunately, the operators succeed in returning the cooling system to normal, and nothing serious occurs. Nevertheless, the nuclear reactor automatically had shut itself down when the situation was approaching a critical point. The plant operation - which signifies electricity production - cannot be resumed before a technical inquiry on what happened in the power station can be made by a federal authority (such severe safety regulations, in force for managing a nuclear plant, cannot be omitted even in a film like this).

However, during the critical moments in the control room, the supervisor-Jack Lemmon has noticed a momentary abnormal vibration of the building, and fears that there is something wrong with one of the pumps of the reactor system. He tries to discuss this with other engineers and the management of the plant, but nobody wants to pay attention to him. The utility cannot afford a prolonged shutdown to make an accurate check and wants to resume electricity production as soon as possible, to avoid losses. Besides, they fear that possible findings in this plant could delay the commissioning of a new one, of similar design, the construction of which is being completed, and this would be a real financial disaster.

The poor supervisor is even unable to explain his doubts to the Commission in charge of the technical inquiry. Incidentally, this is not made clear in the film, so a shadow of suspicion seems to be cast on this federal body, as being more interested in the company's profit than in safety.

Meanwhile, outside the power station, the cameraman is saying that something serious has occurred inside the nuclear plant, and that the management is trying to cover this up. In doing so, he shows the archetypal character of the antinuclearism we mentioned before. Not possessing sufficient evidence to suspect any failure (unlike the supervising engineer), he knows nothing about pumps or possible design faults, nor does he intend to rely on technical information. He has simply looked at the faces of the operators in the control room, and at the lights flashing on the control panel. For him, naturally dressed in casual wear, with long hair and beard, inside that nuclear plant there are only wicked capitalists and their slaves, wearing suits and ties, playing with a risky system and hiding the truth from the people.

The best of all this turns out to be that the wicked really are there, as the engineer is going to discover.

The cameraman, assisted by a Jane Fonda who is quickly becoming conscious of her duty as a reporter and is feeling more and more involved in the struggle, tries unsuccessfully to convince the TV network to broadcast the video he has inconspicuously filmed in the nuclear plant, and which he considers to be conclusive criminal evidence. Meanwhile, news of the plant shutdown is been released by papers and networks. Almost immediately, groups of young people and mothers with children arrive on the scene, demonstrating outside the nuclear plant and demanding loudly how they plan to dispose of nuclear waste (which, frankly speaking, should be the last of their worries under the circumstances!).

The cameraman finds two trustworthy technical experts (so called independent experts appear on the scene). Watching the video, one of them tells him that it is a miracle that they are all still alive, because the reactor has come close to the China Syndrome. What does this signify?

One of the independent experts says that there is a particular type of accident in which the core of a nuclear reactor, having lost cooling and being completely out of control, can melt the concrete base of the building, penetrate into the bowels of the earth, and, with no further obstacle, travel to the opposite side of the world, reappearing in China! But, if the molten core hits a water table, a radioactive cloud may develop, transforming an area as vast as Pennsylvania into a desert.

After such a revelation, the reporter becomes very keen to discover the truth, even against the will of the TV network. She meets the ill-fated supervisor, in a coffee bar close to the nuclear plant. Ill-fated, precisely because of this encounter and for changing sides from evil to good, he is going to lose his life.

In the dark and smoky atmosphere of the coffee bar, Jane Fonda asks him, dramatically, if the power plant is really safe, and if the China Syndrome is possible. The poor man is clearly troubled. He replies that a nuclear power plant is normally safe, that he is confident in nuclear energy, having spent his life at that reactor. He also hints at the possibility of an accident, but insists on one point: a check on the pumps must be carried out in the reactor, and also on those of the new reactor ready to be put into operation.

Days later, the engineer decides to investigate himself, and his findings confirm his suspicions. To save time and money, certain indispensable controls on the welding of the piping system of the reactor had been neglected. He also realizes that someone is trying to murder him! In order to protect himself, and avoid a disaster, armed with a pistol he locks himself in the control room, asking to see the only people in whom he has now confidence, the reporter and the cameraman. When they finally come and begin a live broadcast, the engineer-Jack Lemmon is very agitated and, in the short TV programme, soon interrupted by the wicked, he looks like a hysteric who has lost self-control.

In the end, on one side there are the managers of the company urging the security squad to break down the door and enter the control room, on the other the two network guys watching the scene with impotent rage. Finally, the armed squad breaks in, shooting wildly. The poor engineer dies, without firing.

At first, it is quite easy for the management to declare that a mentally disturbed staff member was seriously endangering the nuclear plant. But soon after, other engineers have a crisis of conscience and decide to tell the truth and make public what had been discovered by their dead colleague. The wicked must then face the punishment that they deserve, whilst the two reporters' merit is finally recognized.

What can we say about this film? Built as a thriller, it does not invite people to reflect on the story. But if someone does, as probably very few did, then a question arises. What is the real problem? Is the nuclear plant unsafe (intrinsically, we would say today), or are the owners of the plant wicked and greedy for money? In other words, is it a message against nuclear energy, or against the cynical capitalist society?

If we have anticipated that this film is a kind of manifesto of the antinuclearism, it is just because it clearly shows, unintentionally for sure, how the two targets – capitalism and nuclear energy – were considered to be equivalent by these groups, and that the concern for environmental protection can scarcely be considered the source of their action and behaviour.

This is also confirmed by the extremely polarized colours used to paint characters and situations, as is typical of propaganda work. It is true that serious accidents like Three Mile Island (TMI) had not occurred yet, but the accident imagined in the film – the China Syndrome – is an example of trashy scientific literature (we cannot speak of science fiction, because the film pretends to be credible). The nuclear engineer himself, who has confidence in nuclear energy, but demands priority for safety, is punished with death, just for this claim.

As for the wicked capitalists, they look more irresponsible than cynical, which is rarely the case for a true capitalist. But on this, the critics of capitalism fall into an embarrassing contradiction, which, however, will only appear evident after Chernobyl, when it became clear on which side of the world cynicism prevailed over nuclear safety.

There is in the film another polemic tool, roughly handled, which has always been one of the most exploited in any antinuclear campaign, especially in the USA, where the memory of the Watergate was still alive: the *cover-up*, which, in this particular case, means willingness to keep people in the dark about the risks of nuclear energy. In a country where the media are not only aggressive but also credible, this would be an error of nuclear energy, able to shake the public much more than the wickedness of the capitalists, and, in reporting it, opponents have never been cautious. The film itself was advertised as being inspired by real but never revealed events, without however providing any evidence, facts or information useful for the public.

A book, published in 1989 in the USA, is a typical example of what opponents have meant by a nuclear accident, and the way that they are used to reporting it. The title is 'The Greenpeace Book of the Nuclear Age', and it shows, among other things, an endless list of events that are presented as nuclear accidents, in which the authors shamelessly mix military events, simple malfunctions of the first experimental reactors and trivial problems encountered in research laboratories. It is worth quoting a few of them, because some are, so to speak, "nuclearly" exhilarating. (It is worth recalling as well that, on the first page, there is a long list of "nuclear experts" – again the famous independent experts we have found in the film - who have worked on the book.)

1959 January 18. An F-100 Super Sabre interceptor aircraft, designed to carry nuclear-capable air-to-air missiles, exploded in flames on the runway of a Pacific base when its external fuel tanks inadvertently jettisoned. (No precise location is given but US bases in the area at that time were in Okinawa, Taiwan, South Korea and Thailand.) The nuclear capsule was not in the vicinity of the aircraft and was not involved in the accident.

1959 July 26. At the AEC's Sodium Reactor Experiment reactor at Santa Barbara, California, a series of test runs revealed that tetralin sealant had leaked into the sodium coolant, where it had decomposed and coated the fuel elements, thus reducing the transfer of heat. Intermittent attempts were made to purge the coolant and clean up the fuel elements.

During the final run there were 10 scrams and four forced shutdowns. When the operators finally shut the reactor down to investigate the cause of these problems, they found that 10 of the 43 fuel assemblies were severely damaged.

1962 March 26. The Royal Navy's first nuclear-powered submarine, the Dreadnought, suffered a fire while *under construction* in Barrow-in-Furness. A second minor fire occurred in the submarine's control room in December 1965 while she was undergoing repairs in Rosyth, Scotland. (Neptune).

1966 October 26. A crewman aboard the aircraft carrier USS Oriskany, operating off Vietnam, panicked when a flare accidentally ignited while being moved, and threw it into a locker containing 650 more flares. The resulting fire took three hours to bring under control, killed 44 people, destroyed or damaged six aircraft and put the aircraft carrier out of action for several months.

1966 November 4. A flash fire occurred on board the aircraft carrier USS Franklin D. Roosevelt off Vietnam. The fire, in a storage compartment containing oil and hydraulic fluid, killed seven crew members. (Neptune)

1967 November 5. The nuclear-powered ballistic missile submarine HMS Repulse *ran aground* 30 minutes after her launch from Barrow-in-Furness in the UK. It took seven tugs to pull her free. (Neptune)

1968 April 9. The Polaris submarine USS *Robert E. Lee* became snagged in the nets of the French trawler Larraine-Bretagne in the Irish Sea.

1970 January 29. The submarine USS Nathanael Greene ran aground in thick fog in Charleston harbour, South Carolina. It took seven hours to refloat the vessel.

1974 January 8. The nuclear attack submarine USS Finback collided with the USS Kittiwake at the Norfolk Naval Base in Virginia, causing minor damage to the hull of the Kittiwake. (Neptune)

1979 March 7. Entangled in the nets of a Scottish trawler in the Sound of Jura off the west coast of Scotland, the nuclear-powered submarine USS Alexander Hamilton towed the fishing vessel backwards for about 45 minutes, until the nets were cut by the trawler's crew.

1981 25 April. Important feedwater cooling systems used to remove heat from the Brunswick 1 and 2 reactors were found to be blocked by oysters, mussels, barnacles and tube worms. On 28

August, at the Pìlgrim plant, a heat exchanger was rendered inoperable due to the growth of mussels in cooling system pipes.

On 9 June (1981). At the San Onofre 1 reactor, the flow of cooling water was slowed and a valve malfunctioned because of an infestation of barnacles in the heat exchanger discharge pipe.

1982 July 19. A fire started by an electrical fault caused damage estimated at several hundred thousand pounds to the new French-built nuclear power station at Koeberg near Cape Town, South Africa. The fire delayed the start-up of the station by some weeks. No one was hurt in the fire.

That's enough, but there is a good deal of similar rubbish in this book. Nevertheless, another of the listed "nuclear accidents" is worthy of mention, since it is perhaps more significant than the others: the accident that occurred in 1987 at Goiana, Brazil, which is well know by true experts of nuclear waste, as it was largely reported in the scientific literature. Owing to a sequence of events that had nothing to do with nuclear energy, a source of Cesium-137, used for cancer radiotherapy, ended up in a disposal landfill of scrap from the metal industry, where it got damaged and a vast area around the site was contaminated. A rather critical situation arose, particularly on account of the difficulty of monitoring the contamination in such a situation. The social alarm in the surrounding area was cynically exploited by environmentalists, who launched a violent campaign against nuclear energy. A complete remediation of the site was finally carried out, thanks to the intervention of true nuclear operators. The management of the Goiana accident has become a classic case of waste management, frequently discussed in international conferences.

It is worth mentioning what two well-known Italian environmentalists (M. Scalia and G. Mattioli) say in the preface to the Italian edition of *The Greenpeace Book of the Nuclear Age*:

> "This book is not a pamphlet against nuclear energy, but a real data base: year by year the accidents that have occurred in nuclear plants are accurately reviewed and explained [...] from such data a sense of rebellion and astonishment is arising [...] towards the arrogance of a small minority willing to keep millions of persons from being informed on events regarding their safety and health..."

Before finishing with this, we take the liberty of a short comment on Greenpeace, whose people, being notoriously fond of sailing around the world, seem obsessed by nuclear propelled craft. Each year, oil tankers leak into the sea about four millions tonnes of oil (a 1980 estimate): have we ever seen those sea guardians noisily campaigning on this?

The Greenpeace book was written in 1989, but the habit of inventing nuclear accidents has been the favourite sport of opponents since their appearance on the scene (the real accidents of TMI and Chernobyl simply accentuated an existing attitude). If, in the military field, a nuclear submarine caught by the nets of a fishing boat becomes a nuclear accident, what happens in the civilian field is even worse, as we have seen with reports about mussels living near the external heat exchangers of power stations, the most conventional and peripheral equipment in a nuclear power plant. Not to mention operational problems of turbines, not a rarity in any electricity generation system, which have been, for years, the most favourite *nuclear accidents* cited by these ideological environmentalists. Since operational problems of this kind weren't reported by the management of the utilities, nor were there press releases on this subject, they claimed that a cover-up was attempted.

One point has to be underlined. Nowadays, in the management of any nuclear installation, from power plants to waste disposal facilities, a so-called full transparency principle is applied, by which whatever occurs in the plant is reported outside, including sometimes purely routine operation. For this purpose a number of dedicated bulletins are quite frequently issued and distributed to the population around the plant. However, this by no means implies that *before* they used to keep everything confidential. Before, by this we mean *before Chernobyl*, the behaviour of the management in giving information was simply normal, like any other technical or industrial organization.

How have nuclear scientists and engineers reacted to the practice of misinformation by opponents? Being used to debate in appropriate venues and to rely on technical arguments or data, they weren't equipped to face groups or individuals talking of devils and fears, and stating technical nonsense. Their answer to the attack has been an attitude of scorn and annoyance in private, while, in public, they simply showed indifference.

The result was a situation in which, first in USA and then in Europe, for several years there were, on one side, opponents speaking of nuclear accidents that had never occurred and continual cover up's, and on the other technical experts who in private felt indignant but in public shook their heads in silence. In between, there were the Safety Authorities, who had no difficulties in establishing the truth, and the Public Authorities, which had to face a different problem. Local government and local politicians cannot ignore alarms and concerns, even when they are ill-founded. Moreover, along with accidents kept secret, another issue, more easily perceived by the public, was being raised by the opponents: the waste produced by the nuclear plants. There was also the media, between the opponents and the technical experts, and they were quite obviously more interested in the screams of the opponents than in the whispers of the scientists and engineers.

This situation began to change with Three Mile Island, but the turning point was Chernobyl. Since then, it has no longer been necessary to invent fantastic accidents, like the China Syndrome, to fight nuclear energy, because misinformation and inventions were attached to the consequences of those accidents. But, above all, what did change was the attitude and psychology of the technical experts, as we anticipated at the beginning of this book and as we will see later.

The China Syndrome is also significant because the concern about nuclear waste was, for the first time, communicated to the general public by using an effective medium. Those mothers with children, and the bearded young people with long hair, shown in the film demonstrating around the nuclear plant and asking for the destiny of waste, will in fact make school in USA. In this sense, the effectiveness of the film as a communication message shouldn't be underestimated, not least thanks to the unexpected success the TMI accident gave it, just after its release.

The problem of nuclear waste, which, since the beginning, had been one of the *leitmotifs* of the antinuclear struggle, is therefore also made public with the help of this film, pushing it beyond the narrow circle of opponents. This is a matter destined, even after Three Mile Island, to meet with the interest of the public in the USA, much more so than the safety of reactors or the maximum credible accident - topics having a truly scientific basis (and which have scarcely been addressed by environmentalists). The subject of wicked capitalists, careless of nuclear safety, also remained quite unfamiliar to the American public, which is, as we know, little inclined to apply ideology in dealing with problems. So did the subject of the claimed 'secret' accidents, primarily on account of the level of credibility in the public institutions, such as those established for nuclear safety, in the USA.

Therefore, whilst the political or ideological claims of the opponents, such as non-transparency or secrecy (included the possible connection with military affairs) and profit having priority over safety, always has remained the arsenal of an excited minority, the subject of nuclear waste has gone far beyond, becoming a banner for those groups. On the other hand, when the opponents of nuclear energy were taking their first steps (in the early seventies), the waste problem could, to a certain extent, still be regarded as unresolved, because at that time the conditioning technology had not yet achieved industrial maturity, at least for some types of waste. Therefore alarm and concern diffused on the subject could not be challenged and proved completely false.

In a society becoming increasingly conscious of the environmental problems that had arisen with progress and industrial growth, the waste issue itself could raise an interest in the general public and also involve families and quiet persons, such as those shown demonstrating in the film around the nuclear plant. As a matter of fact, in the USA, the only widespread concern about nuclear energy was focused on how to dispose of nuclear waste, certainly not on the ideological or political aspects we have seen above. We can even speak of a movement of opinion, effective mainly at the local level in connection with nuclear power plants siting, which never showed any aggressive behaviour.

A movement of opinion based simply on genuine concern, cannot be ignored by national or local politicians, so a legislation was passed making the licensing procedures for nuclear plant more and more difficult and time-consuming (one of the new requirements was, for instance, the obligation to present, along with the application for plant construction, a detailed plan for managing the wastes produced).

This legislation, reinforced after Three Mile Island, made the construction time of a nuclear power station erratic and unpredictable, consequently discouraging the planning of new plants as a risky undertaking. Good use was made of this legislation by the opponents, who found the strategy of legal battles against decisions of public authority and utilities extremely profitable. In this they were helped by the federal jurisdiction of the USA and also by the rather favourable attitude of the courts of justice on this matter.

***

Whilst in USA the political and ideological nature of the opposition to nuclear energy is best shown by Hollywood films, in Italy it can be found in the long and troubled story of the Montalto nuclear power plant, in which the antinuclear groups even joined forces with violent activists of the extreme Italian Left, sometimes equipped with P38s.

Between 1975 and 1978, before TMI, and long before Chernobyl, it took place in Italy what we can call the "tragedy" of nuclear energy, the outcome of which became clear ten years later, with a referendum and the decisions taken by politicians. In a country like Italy, more so even than in Germany, where the struggle also was no joke, the political and ideological character of antinuclear environmentalism was seen to be unconnected with real ecology. It was also emphasized by the peculiar situation of Italy in those years, the years of bloody attacks by some revolutionary groups.

In Summer 1975 a decision was taken to site the second large nuclear power station (the first was Caorso) at Montalto, on the coast of the Tyrrhenian sea, some 100 Km northwest of Rome.

Terrorist misinformation of the antinuclear groups broke out against this planned siting, based on radiological risks associated with the power plant, aimed at raising social concern and alarm in an area where the economy was mainly based on farming and tourism. Irradiated crops, beaches and rivers exposed to contamination, lethal radioactive gases released over a vast territory, nuclear wastes buried somewhere around simply because nobody knew what to do with them: the population of that area south of Tuscany was spared nothing.

This is not an American province, where people use to keep a 'stars and stripes' in their sitting room, and who basically do not distrust public authority. Notwithstanding the developed agriculture and welfare, this is an area that still maintains a rustic character, where people are not easily confident of what the government and its representatives say. Little by little, the incessant campaign of the antinuclear groups proved to be effective with the local population. In 1976, protests began against the Council of Montalto and the regional Council, which both voted in favour of the nuclear plant. Before long, a people's Committee was established which, playing the revolutionary game, tried to face the elected assemblies in the way the *Comité de salut public* and the *Sanculottes* faced the Convention during the French Revolution. There was even a nobleman among the revolutionary leaders (do we remember Mirabeau?), himself a great landowner, promoting railway and roadblocks.

The most involved group, which supplied farmers and noblemen with technical and scientific information, was *Democrazia Proletaria*, an ultra-leftist party which also had among its supporters some scientists – one of these being Marcello Cini, a well-known professor of physics at Rome University – who, instead of providing data and numbers, liked to speak of nuclear imperialism.

As always occurs when there is someone inclined to reason on one side, and on the other someone willing to scare, there was little audience for the real experts coming from ENEL and ENEA, respectively the National Utility and the National Committee for Nuclear Energy. Besides, these ultra leftists had learned one of the best known lessons from Lenin: injure the reputation of the antagonist's party.

Some did this in a brutal manner, openly blaming scientists and engineers as serving the interests of the nuclear lobbies, and others did it in a more sophisticated way, so the nonsense was cryptic and not easily spotted. Let's listen to how, in those circumstances, the aforementioned Marcello Cini, an eminent antinuclear physicist, tried to counteract the arguments of the nuclear technical experts.

> First, we have to admit that any expert is a partisan one. This not necessarily means that he is lying purposely, at least when he is in good faith. This means that experts – as all men do – express opinions which are valid only in a given context, and only this context validates data, scopes and objectives. Everything is valid within its context. But the essential point is that there are many possible contexts: quite often technical experts do not know, or prefer to ignore, what is the context which validates their opinions...They believe, or pretend to believe, that their truth is an absolute truth. For them data are data, numbers are numbers. This is the current ideology in the world of science and technology. This philosophy has to be abandoned, otherwise the future of science and technology is hopeless, because any number can be faced by an other number, valid in a different context...

It sounds like an obscure text, but here's what it means in a more prosaic way: if I say that the nuclear power plant (he was speaking of Montalto) will irradiate the artichokes on the farms nearby, this becomes the truth, because I have put myself in an antinuclear context, not unlike technical experts who give evidence in a pro-nuclear context).

In other words, a return to a rough Aristotelianism, five centuries after Galileo.

Whilst at Montalto the local population was getting worried about crops, fishing and tourism, *Democrazia Proletaria*, the ultra leftist party, was making efforts to extend the protest to a national level and join the antinuclear struggle with a general opposition to capitalist society, thus troubling leftist parties (Socialist and Communist parties) and unions. In the most uncomfortable position was the Communist party, which dominated the elected local councils with a large majority. Since the beginning, it had been in favour of the nuclear plant, keeping a coherent attitude even during the days of the noisiest protests of the opponents. This party thus became the preferred target of the ultra left's criticism, also undergoing serious internal disputes. Communist intellectuals begin to speak of alternative sources of energy "complying more with communism", while others tell their comrades that energy is also needed for developing jobs, and that some intellectuals simply wanted to protect the solitude of their villas in Tuscany. As we will see later, the Communist party of Italy did not forget this lesson of internal conflict, when a decision on national nuclear policy after Chernobyl had to be made.

The open attempt of the ultra left party to turn the case of the Montalto nuclear plant into a political attack against the dominating political parties seemed successful for a while, when the antinuclear opponents marched along with farmers, landowners and WWF activists. At the beginning of 1977, the groups and the People's Committee achieved considerable popularity with the local communities and appeared successful in defeating the plan of ENEL to construct the new nuclear power plant. The other political parties were reflecting on whether to withdraw their support to the nuclear plant. Even the ultra right party joined the field of the opponents, and, to avoid embarrassment for most of them, the protest meetings against the power plant were quite often obliged to end up with mention of antifascism and democracy.

Nevertheless, before long the situation changed rapidly, since the public authority decided to re-establish law and order, helped in this by some initiatives of ENEL, which began to distribute the compensation funds that had been planned by a bill passed in the national Parliament. This gave the political parties the opportunity to emphasize the social advantages of the nuclear plant, thus splitting the front of the opponents and dividing the interests of the local people from those of the antinuclear groups. On the other hand, illegal activity, like road and railway blocks, began again to be prosecuted, after a period of tolerance, and this also contributed to isolating the protesters.

Shortly, only the violent ultra-leftist activists, some so-called metropolitan Indians from several European countries and pacifists equipped with sticks and sleeping bags remained on the field. During the summer of 1977 they attempted to make a rural and seaside replica of the urban guerrillas which, in that year, had troubled several cities in Italy. This frightened the countrymen and farmers of southern Tuscany, who not only definitely dissociated themselves from the groups, but even pushed the opponents, sometimes quite rudely, to leave the area.

In 1978, earth-moving machines were again working on the site for the plant construction, which was since then carried out without major opposition.

Nothing can explain the Montalto antinuclear struggle better than the words used by MM Scalia and Mattioli, two leading Italian Greens, who traced out, in an article issued in 1978 by an environmental magazine, the lesson learnt from the story and from the defeat of the groups at Montalto.

> The expectation of social changes arisen with progress of leftist parties have been frustrated in this country because of weak political opposition carried out by those parties [...]. People threatened by huge nuclear plants, willing to decide about their own lives and their environment, are frustrated by the authoritarian face of the State [...]. These are two aspects of the same political picture. Antinuclear movement has to be regarded as an expression of national political movement originated by social marginalization of the working class, and the refusal of nuclear power plants as a refusal of a society controlled by capitalism's iron hand [...]. If in other European countries the struggle of antinuclear movement has a rather ecological character, aimed at preventing pollution and negative environmental impact, in Italy our objective is to fight the

> interests of multinational Companies and those of national industries allied with them [...].

That's enough, we believe, to realize to what extent environmental protection, which means the protection of nature and of human beings from pollution and degradation, was a worry for the Italian ecologists of the seventies.

Unfortunately for the country, when the leaders of the antinuclear movements drew this conclusion from that defeat, they did not know that Montalto had been a lost battle of a victorious war. The political ruling class of Italy learned from that story the only possible lesson that could be learnt in the country, which was evident after Chernobyl. After that event, worried about a possible new struggle of a noisy and violent minority, and being afraid of losing political consensus, they unanimously decided to finish with nuclear energy. This was done without trying to explain what had happened, without any consideration about energy consumption and supply, without any concern for the huge amount of wasted money. But for one of the Italian political parties there was something more, as we will see later.

***

The Three Mile Island accident, which occurred on March 29, 1979, had no serious effects on people and the environment, but proved to be beneficial for the nuclear reactor safety and for Columbia Pictures, producer of *The China Syndrome*, which, thanks to that event occurring apparently at the right moment, made 26 million dollars with the film.

Let's recall briefly the dynamics of the accident. The so-called starting event was a failure of a pump in the secondary cooling circuit, supplying steam to the turbine, in one of the two units of the power station (TMI 2). This can be considered as a typical malfunction, since it is one of those always envisaged at the plant design stage, in order to show that, should it occur, dangerous effects would be prevented thanks to a duplicate system (sometimes more than one) automatically activated, able to ensure cooling of the core in an emergency. Due to some mistakes by the plant operators (which are not detailed here), the emergency cooling circuit was shut-off, leading to a sudden and uncontrolled increase of the temperature of the core. The nuclear fuel was almost completely damaged, part of it even melted, and fell to the bottom of the reactor vessel, making a mass of ceramic and metal. The core of the reactor was destroyed, and the event, the most severe accident that had ever occurred in a nuclear plant, had been until then considered *almost* impossible.

Nevertheless, only some gaseous short-lived radioactive isotopes, a very small fraction of the radionuclides in the nuclear fuel, escaped to the environment, through the stack. The remaining radioactive inventory was kept inside the containment. In other words, the principle of the so-called "defence in depth" proved to be effective, even in an event that should never have occurred. Or, to be more scientific, which had a remote probability of occurrence.

The most serious problems arose with the management of the external emergency, especially because of the way the population was informed. The State authorities hesitated for a long time about what to do, conscious that some necessary precautions had to be taken, but also unwilling to cause undue social alarm and, perhaps, panic. Finally, two days later, they ended up by deciding to evacuate pregnant women and children from the area surrounding the plant, as precautionary measure. A decision, seen with a coolness they lacked at that time, that was clearly either late or useless (and, under the circumstances, useless).

However, if the defence in depth system, common to all the nuclear reactors of USA design, worked out (a system, as we will see shortly, not existing in the Chernobyl-type reactors), the TMI accident showed some serious problems which led to the revision of three aspects regarding the safety of all nuclear reactors designed and operated in the western world. The first were the criteria for evaluating, at the design stage, the risk of a severe accident in a nuclear power plant. These design criteria, established early in a well known technical Report issued by the Nuclear Regulatory Commission, also known as the Rassmussen Report, has since been reconsidered by adopting a much less optimistic approach. The second was the so-called human factor: in the design of the systems planned for ensuring safety in case of malfunction, after TMI much more attention has been paid to the possibility of wrong operation by the technical staff of a nuclear plant. Finally, the procedures planned for facing a nuclear plant accident have been deeply revised and upgraded, with particular emphasis given to those directed to managing the external emergency. This was probably the most evident lesson learnt by the TMI accident, and since then the preparedness of an *external emergency plan,* also involving local authorities and the public, is a major duty of designers and operators of any nuclear facility.

Whilst science and technology learned an important lesson from the TMI accident, the antinuclear environmentalists launched a wild campaign about the effects of the accident. For about a decade, although the federal and the Pennsylvania State authorities continued to provide quite reassuring data on the radioactivity release and the radiological effects, the opponents incessantly turned out data and numbers, supplied by their "independent" experts, showing incredible destruction of lives. The less disastrous estimate is of about 50.000 *excess deaths* for the period 1980-82 alone, in the affected area (that they considered extended to most of the USA!), not to speak of the claimed contamination of the Susquehanna River, along with the whole hydrographical basin of Pennsylvania, because of radioactive discharges that were kept secret. These are estimates and figures of the same kind as those (like a fire during the construction of a nuclear submarine) that we have quoted from the *Greenpeace Book of Nuclear Age.*

Yet, in 1990, a book of the same kind was published in the USA where, among other things, even the origin of AIDS was connected to radiation from nuclear accidents. The title? *Deadly deceit. Low level radiation, high level cover-up.*

By no means do we want to assert that nuclear energy is absolutely safe - this isn't the purpose of this book - especially when considering the possibility of severe accidents involving power plants. Nor do we want to assert that we alone have access to reliable and true data. We simply want to affirm that claiming disasters of that size, in such evident contradiction to official data provided by public authorities (and not by Metropolitan Edison, the owner of TMI plant), goes far beyond simply claiming a case of cover-up. It implies that all public institutions deliberately aim to deceive people, even when public health is involved, and, moreover, that they do this with the help of the media (which in fact have never taken those claims seriously).

Those, among the antinuclear environmentalists, less inclined to claim disaster associated with TMI, have emphasized, in order to stress the inconvenience of nuclear energy, the technical and financial problems to be faced in remediation of the damaged TMI unit, where the whole core was reduced to a mass of highly radioactive debris. This is, of course, a different matter. Dealing with decontamination, conditioning, transportation and storage of radioactive material and damaged equipment, that were produced because of the accident, has been a very complex and difficult task, which has also been the spur for development of remote handling and transport techniques. There has been nothing technically difficult to overcome, but the whole process has been extremely expensive (for the American tax payers too, because part of the costs for remediation and waste disposal has been covered by the federal government).

The financial impact of the Three Mile Island accident has been an important subject, not only because of its obvious consequences for the assets of Metropolitan Edison, but also for its long term effects on the nuclear industry's economy, due to the new and more stringent safety criteria for design and operation of nuclear power plant, that were introduced soon after the accident. The very high cost increase incurred by the utilities, which also involved power stations already in operation, only took a few years to make nuclear electricity uncompetitive and to discourage any new investments in nuclear power plant construction in the USA. This is further proof of how, in nuclear energy, real priority is given to safety, not to profit.

Among those who have taken advantage of the TMI accident, we haven't listed the antinuclear movement, although it gained in visibility from the event, at least in the USA. Unlike the Chernobyl accident, the event, which was not at all environmentally or socially dramatic, did not have a major impact on people and public opinion not directly involved, and, above all, it did not spoil the technical experts. In other words, it was not suited as the basis for an assault on nuclear energy.

Instead, the event was the occasion of a technical debate, involving scientists and engineers, mainly focused on the criteria used for the probability assessment of severe accidents, established in the Rassmussen Report mentioned earlier, which had been taken as a Gospel until then. These are, as we can see, subjects for true experts, definitely not suitable for noisy campaigners, who could only focus on the claimed effects of the accident, a matter where statistics and figures, as we recalled above, could be readily manipulated.

But there is something more: for true technical experts, the accident itself has proved, *e contrario*, as ancient Latins said, the efficiency of barriers provided by the defence-in-depth system of nuclear power plants. These barriers proved able to prevent significant radiological releases to the environment, even in the case of a complete loss of cooling - an accident that was considered almost impossible.

This is why the self-confidence of technical experts was not seriously shaken by what occurred at TMI. Much more, the accident shook the nuclear industry, as we have seen, and this is also something to think over. After Three Mile Island, it never occurred to scientists and engineers to take into account the economic worries of nuclear entrepreneurs and plant owners, quite often their own employers, and to try and relieve the impacts of new design and operation criteria - nor did scientists and engineers consider them excessively binding. This again shows how little nuclear experts have been responsive to lobby interests, despite what the antinuclear activists, always short of sound arguments, have never ceased to hint.

The ideological environmentalists, then, couldn't profit from TMI to bring nuclear plant safety to trial, a matter technically difficult to deal with, and even risky, because the accident itself could be seen as evidence for the defence of nuclear safety. Instead, notwithstanding an accident whose seriousness cannot be understimated, in the years afterwards, radioactive waste disposal continued to be the preferred battlefield for the opponents. Disposal is, among all nuclear activities, the one without major risks, as we will try to explain in some detail in the next chapter. However, it is also a field where, when technically and scientifically sound arguments cannot be advanced, one can evoke situations and scenarios concerning centuries and millennia yet to come.

In the years following Three Mile Island, the first efforts to select suitable sites for nuclear waste disposal began in several countries. Against these programmes, the antinuclear activists launched campaigns even more virulent than those against nuclear power plant siting, making waste disposal the number one problem of nuclear energy. Pretty soon, they actually identified in the lack of a "final solution" the strongest argument for challenging nuclear energy and the most fertile ground for spreading public doubts and fears. Therefore, to intercept and disturb programmes directed to select suitable sites for waste disposal became, after TMI, a priority for antinuclear activists. Even if it sounds like a paradox, after the accident, the problem of nuclear power plant safety – actually, *the problem* of nuclear safety – has yielded to that of the final disposal of waste, which since then has been claimed as the real Achille's heel of nuclear energy.

The TMI accident not only failed to depress scientists and engineers, it also did not significantly change the perception of nuclear energy by the public, particularly in Europe. This is also shown by the story of the Italian power plant of Montalto, that we recalled above. The plant construction, which was under way when the accident occurred, was in no way disturbed, despite the battle carried out just a year before by antinuclear activists. Indeed, the strong ideological character of the antinuclear struggle prevented the movement from influencing public opinion, or national programmes on nuclear energy.

Everything changed with Chernobyl.

***

When some vague news about a possible accident in a nuclear plant in Ukraine were first reported by the western newspapers – the evening editions of 26th April, 1986 – a town with 45,000 inhabitants had already been evacuated, in an exodus whose nature and scale were perhaps more than biblical, and which only a disciplined and authoritative community like Soviet Union could achieve in a few hours, during the afternoon of April 27th.

From this disaster, however, that hit the population of Ukraine, Belarus and Russia so hard and which devastated the image of nuclear energy, science and technology have learned little or nothing, nor has anything even been learned about how to manage a nuclear emergency. Indeed, the incredible sequence of carelessness, lack of safety precautions, lack of coordination and information, missed compliance with safety procedures, can by no stretch of the imagination be regarded as something from which science and technology progress.

This devastating sequence of operational deficiencies took place in a reactor having *by design* intrinsically poor safety. It is well known that the Soviet designed RBMK reactors have some practical advantages – mainly concerned with construction and operating costs and fuel core management – and also a serious weakness on the safety side. Not by chance, at the beginning of commercial power plant development, this type of reactor was abandoned by the western countries.

Without going into details of reactor physics – which are indeed rather important in this case - we may say that the RBMK reactors shared many of those technical features that make "convenient" a reactor, but at the same time create difficulties in maintaining nuclear and thermal stability. The core arrangement was of the channel type (implying a simplicity of design and construction), the moderator was graphite (a relatively cheap material, but not compatible with water) and boiling light water was used to remove heat, with steam being fed directly to the turbines (a feature upgrading the efficiency of the reactor). Besides, the core pressure tubes were of simplified design (equipped with a single cladding) and, in order to increase the power output, they were densely arranged, a situation increasing the potential instability of the reactor, thus requiring delicate operations with the control rods. Last, but not least, the core arrangement and other systems incorporated in the reactor design led to a huge reactor building, in order to get a power station of sufficiently large capacity, as required by the Soviet programmes: this in turn made it impossible to provide the reactor with an external confinement barrier.

In western reactor designs, where one of the convenient features mentioned above is incorporated, it is always implemented in a much more reliable way. The Canadian CANDU power plant, for instance, where a channel type core is employed, applies a completely different design, including double cladding for the pressure tubes and heavy water as moderator and coolant. The British AGR power plants are moderated by graphite, but they use gas as the coolant, not water. Furthermore, as is well known, western reactors are provided with an external confinement, a concrete barrier completely surrounding the so-called nuclear island (which is the barrier that prevented the escape of radioactive materials from the TMI plant). Moreover, the RBMK power plants were not equipped with a system to treat off-gas routinely released through the stack, which occasionally contains some short-lived radionuclides coming from the cooling circuits. Consequently, these reactors released to the atmosphere, even under normal operations, an amount of radioactive gases that would have forced any western nuclear plant to close down.

In conclusion: the RBMK reactors were, if a joke is allowed even in tragedy, a real fortune. They had been designed in the Soviet Union (along with a number of pressurized water reactors) in the seventies, when the confrontation with the western world was more acute. At that time, the strong need for oil export due to economical problems, made necessary the rapid installation of nuclear power plants, using the available technology.

This was the technical and operational background to the Chernobyl disaster. As for the sequence, here we can just outline the main events leading to the accident.

Events commenced with a planned shutdown for routine maintenance of Unit 4 of the power plant, on 25 April 1986. It was decided to profit from the shutdown to mount an exercise to verify whether, in the case of a loss of station power, the slowing turbine could keep providing enough electrical power to operate the cooling water circulation pump and other emergency equipment, until the emergency diesel generators began to supply power. During the experiment, the lack of coordination and information between the reactor operational staff, the team charged with the test and the electrical load dispatcher (who unexpectedly interrupted the shutdown because power was needed for the grid), triggered off a sequence of misunderstandings and operational errors which brought the reactor to a situation of instability, inevitable from the basic physics of the RBMK reactor.

For about two hours (an eternity indeed, under the circumstances), the operators tried to achieve stable conditions, but it was then that the inadequacy of the staff, the poor planning of the experiment along with a lax attitude in implementing safety procedures, emerged dramatically. A couple of examples: the control rod system, for the RBMK's much more than for other reactors a crucial one, had not been operating as established by standard written orders; the automatic trip which should have shut down the reactor in case of low steam pressure had been circumvented.

Several times during those long two hours, stable conditions were about to be achieved, but each time an error of the operators occurred or a technical specification was not properly carried out. Finally, the decision taken by the operators to reduce the coolant water flow, presumably to maintain the steam pressure, was fatal, because, simultaneously, the pumps powered by the slowing turbine were also providing less cooling water (as planned by the experiment, which was still going on despite everything). The loss of coolant, by increasing the steam production, caused the definitive transition of the reactor to an unstable condition (as clearly predicted by theory). An overwhelming power surge – about 100 times the normal output – and the consequent sudden temperature increase ruptured part of the fuel and the channels.

What happened – at 01.23 a.m. on Saturday, 26th April 1986 - was a chemical, not a nuclear explosion, which was due to the violent chemical reaction of hot fuel particles with water (a so called *steam explosion*). This explosion destroyed the core, while a few seconds later a second one, also chemical in nature but probably due to hydrogen, brought down the roof of the building. The plume of radioactive fission products, debris and smoke from the building rose up to 1.5 km into the air. The heavier debris in the plume, containing most of the radioactivity released, were deposited within a few km's of the reactor, a zone including Pripyat (the town evacuated the day after). The lighter components, comprised mostly of fission products and the whole noble gas inventory, were blown to Belarus and Russia, where vast territories were contaminated.

A particularly tragic chapter, causing the greatest losses, was the graphite fire (31 firemen were killed), which started at 5 a.m. the same day, just when the conventional fires had been put out, and which continued until May 9th. Very little was known at that time about how to face graphite fires of that magnitude, and there was concern that using the wrong materials could result in a further dispersion of radionuclides or even provoke a criticality excursion in the remaining nuclear fuel. From helicopters, 5000 tonnes of various material were dumped, each designed to fight a different feature of the fire. These included sand, clay, boron compounds, dolomite and even lead. Knowledge on the real effects of some of this material – whether beneficial or negative - is still speculative today.

By May 9th, once the graphite fire had finally been extinguished, work began on construction of an emergency containment, which included the famous Sarcophagus, a 300,000 tonne steel and concrete structure, which still entombs the destroyed reactor. This work, continued up to 1990, employed some 600,000 people – the so called *liquidators*.

Besides the town of Pripyat, all the people living within 30 of the reactor were gradually evacuated, bringing the total evacuees to about 135,000. Today, 2003, there is still an exclusion zone of 30 km around the reactor site, fenced and with military surveillance, wherein some hundred radioactive waste repositories have been created.

It is not possible here to go over everything that happened after the accident, from how information was released by the Soviet authorities, to how European countries reacted; from how the "Chernobyl cloud" traveled around the northern hemisphere, to how the event was covered by the media. Not to speak of the huge financial burden, which mostly lies on the shoulders of western tax payers, for remediation of the affected areas and upgrading the safety of RBMK reactors that continued operation after the accident. Each one of those subjects might deserve a book, and, in fact, volumes have been written on the Chernobyl accident, in addition to the innumerable articles published in newspapers and magazines. Just to give an idea of what happened inside the country, we might simply recall that the accident gave a not minor boost to the dissolution of the Soviet Union, although Gorbachov desperately tried to reassert on that occasion the famous principles of *glasnost* and *perestroika*.

Instead, we want to understand and explain how and why Chernobyl had such psychological and sociological effects, and in particular it is worth looking closely at two points, crucial in the history of nuclear energy: how the accident was perceived in western nations (and the role played in this by the antinuclear activists and others), and the discouragement it spread within the scientific community. From the combined effect of these two factors, arose what is nowadays recognized as the problem of the social, or more precisely sociological, acceptance of nuclear energy - a problem, as we have anticipated, having today an essential influence on nuclear waste disposal.

Before Chernobyl, nuclear dread and demonization, scientific nonsense and other stuff of this kind about nuclear energy, were arguments belonging to small groups of excited people, similar to the cameraman we found in *The China Syndrome*, or to activists struggling against bourgeois society that we have seen at work in Italy, around Montalto. They were part of protest movements challenging "the system", which had no significant audience in public opinion in the democratic countries, notwithstanding a worried perception about the uses of nuclear energy that had been growing among the public, as we saw in the previous chapter, mainly on account of the claimed connections of its peaceful and military uses.

After Chernobyl, nuclear opponents did not change their scientific weaponry, nor their polemic tools; rather, as after Three Mile Island, they worked - and still are working - on spreading terror about the consequences. This time, however, they will find many more people prepared to listen to them.

It took several months to reconstruct the sequence of events we have described above, while the actual thermodynamic development of the accident has been under technical and scientific discussion for years. On the one hand, we can consider it quite normal that the understanding of such an event required time to come out, also because, in the very short term, other problems had clearly to be faced. On the other hand, what happened with information is perfectly consistent with the Soviet system. It was, indeed, a "classic" of the Soviet Union.

With Chernobyl, the most striking and maladroit case of *cover-up* of recent history took place. Maladroit indeed, since the Soviet authorities kept on releasing sparse information and even denying that any serious accident had occurred, while in several European countries, beginning in Scandinavia, abnormal atmospheric contamination had been recorded and measures were already being taken. This behaviour significantly facilitated the unbelievable *debacle* that overtook Europe – from the media and the public through to the politicians.

The contamination recorded by the European monitoring systems was soon considered as an emergency – despite the fact that the data registered did not call for such a reaction outside the Soviet Union – under the pressure of three factors (all considerably emphasized by the lack of reliable information). First, the existing legislation, completely inadequate at that time to face the problem of transnational contamination; second, the attitude taken by the media, which helped more to alarm people than to provide correct information; finally, the immediate and noisy entry into the arena of the antinuclear activists. These factors, much more than actual measurements of radioactivity, much more than data on airborne contamination exchanged by international organizations, influenced the decisions made by national governments. None of the measures taken came from coordination or consultation among countries; each country decided and acted on its own. If simple and harmless actions and countermeasures, such as the recommendation to wash fresh fruits and vegetables or to refrain from using rain water, were more or less shared by several countries, others, that were less innocuous and had more impact on day to day habits, were completely uncoordinated and, quite often, conflicting.

One example will be sufficient – that of milk, the consumption of which, being an essential nourishment for children, was the most subject to assessment and intervention. At the beginning of May, the government of the Netherlands set a contamination limit of 500 Becquerel per litre for milk (an extremely low concentration), and 175,000 litres were confiscated. In Belgium, the government said milk was safe for consumption, but also recommended caution in giving it to children (!). In Luxembourg, the public authorities claimed milk to be perfectly safe for both adults and children.

The response of the quiet Benelux countries can perhaps be better understood if we keep in mind that in the Netherlands there was one of the most active and ideological antinuclear movements of Europe, while in the other two countries opposition to nuclear energy was negligible. In the Federal Republic of Germany, the country with the most violent antinuclear groups, conflicting measures and recommendations between *Länder* (especially when governed by different political coalitions) and federal government led many near to panic. In Munich, the Greens suggested evacuating small children to Portugal (a country considered unaffected by the Chernobyl cloud), and some pregnant women even had abortions, scared of possible effects on their children. A poll showed that 50% of people thought that the government had tried to hide the consequences for public health. The country's entire stock of iodine tablets was sold out by April 30, while on May 6, no more Geiger counters – an instrument for radiation detection and measurement – could be found anywhere in Germany, even though they were not suitable for air contamination detection.

In Italy, the sale of fresh vegetables was banned, which was also a rather more political than health decision. Had they advised to wash fresh vegetables more carefully, as the public authorities did in other European countries, then, as well as removing possible traces of radioactive Caesium, the copper associated with chemicals routinely used to make salad leaves green and bright, would also have been washed off.

For some weeks, even the Common Market in foodstuffs within European Community crumbled, although free circulation of goods was at that time the only cement of Europe, enshrined in the EC Treaty. Not only did some Member States set restrictions on the import of food coming from another State that *could* have been subject to contamination, but each one chose to fix its own contamination limit. Some governments even set contamination thresholds for imported goods lower than those they applied within their own country. In addition, nations with no significant nuclear power production, governed by political coalitions including environmentalists, such as Austria and the Netherlands, decided to stop the import of some foodstuffs, unless a zero level of radioactivity would be certified, which is almost impossible, on account of naturally occurring radioisotopes. Several months were required to re-establish a minimum of common ground, while much more time was needed to normalize the circulation of produce within the European Community.

If differences across national borders might appear somehow logical, since they reflected specific lifestyles or food habits (such as the restriction on dairy consumption, taken in Austria and Switzerland, but not in the Mediterranean countries), others appeared purposely directed to raising concerns or, worse, distrust among people. This is what happened in those countries where the public authorities, having first reassured people on the basis of collected radiometric data, then - under pressure from the mass media and anti-nuclear activists – began to give advice on how to avoid risk (such is the case of milk in Belgium, recalled above).

It was not easy for the democratic nations to face the Chernobyl cloud and take logical and rational measures, solely aimed at protecting public health. But above all, between the mass media, fostering alarmism and blaming the authorities for inactivity, and the opponents, launching a campaign about the dreadful consequences, the scientists and engineers were in particular difficulty.

For ten days after the accident, they could only work on data supplied by the monitoring networks, while, about the accident, they could simply try to understand what had happened by interpreting unreliable press releases and information provided by the Soviet government. This was the moment for the experts in radioprotection, whose performance did not prove to be reassuring. Used to being divided into hawks and doves, which in practice means being for or against nuclear energy – we are simplifying, since it's not possible here to open a discussion on radioprotection and low doses – they did not help to make possible a common position on contamination thresholds and radioactivity limits in foodstuffs.

Finally, it began to be realized that what had happened was the destruction of a large commercial power plant. While the exultance of the antinuclear activists was soaring (exultance is not an inappropriate word: those who went through that experience know that, in some circles, this was the dominant feeling), the first to take the floor were the experts of the national nuclear safety authorities. They were – and are – true experts in nuclear energy. Unfortunately they paid for their lack of training and poor attitude on communication, as later did other experts from commercial companies and national agencies.

We had seen the technical experts maintaining a dignified silence, or whispering, while the opponents were making communication their most effective weapon, spreading unrestrained misinformation and scientific nonsense. While it is not true that nuclear scientists and engineers have carried out their activities without transparency - as it was claimed before and after Chernobyl – they have completely disregarded communication, even when, with the ideological environmentalists coming to the scene, it had become a clear need. It would be incorrect to blame nuclear experts and organizations for not having reacted adequately with counter-propaganda. Nevertheless, they could be reproached for being narrow-minded, if not intellectually haughty, for not having envisaged, even after Three Mile Island, at least some public relations activity.

Any way, the task of experts from the national safety authorities of western Europe of giving information and providing data, in the middle of a claimed nuclear emergency and under the noisy pressure of anti-nuclear groups, was not only difficult; it was uncomfortable.

Difficult, because reliable information on what had happened was not available and the radioprotection experts were not supplying sound and uniform guidance: on the contrary, some trade barriers between European nations were suddenly being erected, as we have seen. Uncomfortable, since, having to play the role of national guardians of nuclear safety, they could not easily give reassuring information without exposing themselves to the risk of being accused of minimizing the event or underestimating its effects. If we also keep in mind that several technical experts that took part in TV debates and round tables - interminable in those days – turned out to be absolutely unsuited for communication, we can imagine the success with the media of those who tried to explain quietly and to be rational.

We must say, however, that reassuring people and giving reasonable advice was then an almost impossible task, particularly in Italy. There were mothers who did not know whether the milk was safe, pregnant women incessantly finding newspapers articles on the effects of radiation on the foetus, farmers obliged to destroy their crops. Finally, there were antinuclear activists claiming everywhere that the radiometric data supplied by the national authorities were not reliable, because they came from *nuclear lobbies*. In other countries, like France and the UK, the excitement of public opinion was prevented by a greater *sang-froid* shown by both government and mass media: a behaviour considered by two well-know anti-nuclear activists in Italy – MM. Scalia and Mattioli – as proof of: "[...] *nuclear blindness, a worrying mass regression, involving information as well as democracy* [...]"

In the democratic nations, where for months reports on actual or presumed Chernobyl effects had been dominating the media, it was inevitable that public opinion was to a greater extent influenced by unrestrained opponents – continuously crying "*we told you so!*" – than by technical experts quietly giving advice. For these, the crisis was dramatic, and many fell silent. For people raised in the school of nuclear safety and used to operate under its criteria, a commercial reactor blowing up was not merely an event to look at scientifically. For many of them, the accident was a bitter disappointment and the collapse of their genuine technical convictions. Besides, only an inner circle of nuclear experts was familiar at that time with the

RBMK reactors, while most were simply acquainted with their operating principle, as reported in the reactor handbooks. The lack of containment barriers (which is not necessarily an intrinsic reactor feature but rather an implementation option), the lack of acceptable safety precautions, along with the missed compliance with operational procedures, all came out, little by little, in the following months, sometimes in seminars or technical workshops on the accident. These occasions, as we can imagine, are scarcely capable of raising the public interest, particularly when, under the circumstances, the mass media were focused on the tragic events involving the population, with towns to evacuate, thousands of children requiring assistance and western countries urged to provide help.

This is how Chernobyl became, in the public perception, the sign and the evidence of the danger associated with nuclear energy. Since then, this energy has been found guilty, splashed across the front page, of a crime for which the responsibility lay elsewhere, as will be seen later.

With technical experts reduced to silence and suffering a real intellectual crisis, the post-Chernobyl period was the great moment of the antinuclear groups. Clearly disappointed by the Three Mile Island accident, they now found an event as great as their ambition. To begin with, they tried to raise again, on a larger scale, the connection of peaceful and military uses of nuclear energy, by hinting that an experiment was under way in the destroyed reactor, to generate weapons grade plutonium (which is technical nonsense for a country like Soviet Union, as we explained above). They continued by revisiting the TMI accident, telling nonsense about a huge hydrogen bubble which, just by chance, did not explode, so that only by a miracle was a disaster avoided. Finally, the antinuclear activists attacked – and never have ceased to do so - the official figures reported about the Chernobyl effects, claiming that they are largely (and purposely) underestimated. Those who relied on data and measurements, included radioprotection experts, were sentenced without appeal as the accomplices of governments and nuclear organizations in hiding the truth from people. With this, a real campaign began directed at shaming nuclear institutions, safety authorities included, which has led to serious and lasting consequences. The present day problem of the social acceptance of nuclear energy, the effect of which weighs heavily on waste disposal, as we know, arises primarily from the *credibility gap* between people and institutions, a direct outcome of that campaign. (Why, in fact, a safety assessment carried out by an independent national Safety Authority, is today considered not sufficient for convince the public that a repository is safe?)

With Chernobyl, someone among these environmentalists went so far as to criticize the western world – an incredible performance indeed. Let's listen again to what MM. Scalia and Mattioli, our Italian Greens, said soon after the accident:

> ...the safety of the western reactors is compromised by the obvious consideration that, coming more or less directly from military technology, their design did comply with the interest of the military-industrial complex; this means that need for commercialization has had priority over safety, in order to repay the huge cost incurred......

Such an incredible statement, much more than other declarations made by the antinuclear activists after TMI and Chernobyl, gives an idea of what they dared to say. But to fully appraise their impudence – and their intellectual dishonesty - it is necessary to stop for a while and to consider what were the actual causes, and the actual effects, of Chernobyl.

***

Several nuclear power plants in the western world lie in the middle of farming belts, agricultural areas with scattered hamlets, well-kept farmhouses, irrigation canals, where one can see at work modern equipment and machinery. From northern Italy to the Rhône valley, from Cornwall to New England, the landscape is more or less of this kind, showing two distinctive faces of western civilization – developed agriculture and advanced technology.

When we drive along the eighty kilometers separating Kiev, Ukraine's capital, from Chernobyl, as the urban environment fades away we run into and recognize situations and features that Chekov depicted more than one hundred years ago and, before him, the Ukrainian Gogol. Ox-carts, poor *isbas* with rickety fences and chickens scratching around, countrywomen walking by the roadside wearing a cloth square as headdress, and using another for a bag. Whilst, in the former Soviet Union, some industrial districts – like those existing around Kiev – are striking for their dreariness and degradation, this countryside impresses with its unchangeableness which, far from appearing bucolic, represents the pre-industrial rural character of the nineteenth century, even though it is one of the most important areas of Ukraine – and previously of the Soviet Union – for agricultural production.

Now that we know what happened at Chernobyl, both the power plant and the surrounding environment can be considered, not unlike in western countries, as two distinctive faces of a single system, of its organization, and even of its scale of values.

After the accident a debate took place, in scientific conferences and among international agencies, on whether deficiencies in reactor design or in operation procedures had to be called to account for the event. This debate wasn't totally free; that is, it was not free from prejudice and non technical constraints, especially in the most eminent international organization, the International Atomic Energy Agency, of the United Nations. Actually, while the accident sequence has been reconstructed almost minute by minute, stimulating an extensive scientific literature, the issue of the real causes of the event has continuously been sidestepped, while no word has been said openly about the responsibility, even though everyone knows it.

A technical inquiry has been carried out that can be compared to a manslaughter trial, in which the details of the crime are perfectly known, the weapon has been found, along with the way it has been used, but where the person responsible for the crime is not identified, because everyone is reluctant to mention its name. However, for non-naturally occurring disasters, there is always someone or some group responsible.

Why did this happen?

It was understood quite early by technical experts that what had happened in the Soviet nuclear plant was the product of an almost total lack of "safety culture". A culture that, in western countries, has always dominated the peaceful use of nuclear energy, and in particular the sector of commercial reactors, leading to an unequalled level of safety, which in modern industry is perhaps shared only by aviation. Instead, the lack of this safety culture is visible in every stage of the Chernobyl story: in reactor type and plant capacity selection, in establishing operation procedures, in selection and training of personnel.

The safety culture, as a general criteria for technological application, requires for its implementation two conditions. The first, a fundamental one, is not of a technical or economic nature; it deals rather with the notion of both human rights and environment, and with the instruments a society has devised to protect them. In other words, it is democracy, without which safety culture is meaningless and not applicable, since independent controls wouldn't be allowed, included the indirect control made by public opinion. The second condition is an acceptable level of quality in industrial organization and management, which in turn cannot be achieved without at least a minimum of personal motivation.

In a totalitarian regime, the first condition is obviously not met, since absolute priority is given to the interest of the State, and in the USSR the interest was to build nuclear power plants, to do it quickly, using the available technology and minimizing costs. But in the Soviet Union the second condition, namely a decent level of organization in industrial activities, was also absolutely not satisfied, since an impressive level of carelessness and a general lack of personal commitment were the dominant features of that society. This is indeed a "boundary condition" that nuclear energy cannot afford at any stage of its development and in any field of application. Such condition is in fact prevented, even in less developed societies, when dealing with nuclear energy. But in the Soviet Union the mix of these two conditions – a totalitarian regime and carelessness – was deadly, and this is the real background of the Chernobyl accident.

Therefore, if we want to be plain, since we are not under any obligation of political correctness, like the international organizations, we can say that Chernobyl is by no means the son of nuclear energy, but a legitimate son of soviet communism.

No one has made a film about Chernobyl, nor has there been a *China Syndrome 2*, and the antinuclear "democratic" fellows from Hollywood have vanished. However, it would have been better to resume the argument, since, on closer inspection, there was much of Chernobyl in that Californian nuclear power station imagined in the film. There was, first of all, the need for electricity production to the detriment of safety; there was the cover-up, there was even an armed and authoritarian management inside the plant. Interesting enough, there was also a nuclear engineer who, as he wanted to give priority to safety, ends up by being killed. An example - perhaps too bloody (after all, the more silent Siberia solution was available) - of how a dissenter could be treated in that country.

To hear, after Chernobyl, what has been said about western reactor safety by the two Italian antinuclear activists aforementioned, one has to believe that, instead of trying to understand what had happened, they hurried to see *The China Syndrome* again, without noticing that the story belonged to another continent.

Unfortunately, the true story of Chernobyl has been, if not hidden, darkened. As the truth emerged little by little, the first months afterwards was darkened by the dramatic news concerning the local population, and later by the frightening campaign of the antinuclear activists on the consequences – a campaign that is still going on. Finally, the truth on Chernobyl has been somewhere darkened by politics, as it will be seen. And in all these circumstances, the technical experts did not say a word, at least not in public.

The consequences of Chernobyl, particularly the number of people who died in the accident and afterwards, are an issue where the disinformation disseminated by antinuclear activists has worked extensively and without restraint, since they have, as we know, their own special way of accounting for causes of death.

We do not devalue the awful human and social costs of that tragedy, nor do we underestimate the environmental devastation of large areas around the site and in the three countries of the former Soviet Union. Here, we simply intend to supply actual data and figures, namely those provided by international institutions, including the United Nations, the World Health Organization, the OECD, and by scientific Institutes and Universities, among which the most active have been those in Scandinavia.

Seventeen years after the event, it is possible to make a rather accurate assessment of the radiological and health impacts of the accident and to evaluate the expected future consequences.

The release of radioactive material due to the accident is estimated to be from 100 to 1000 times lower than that due to all the nuclear weapon experiments carried out in the atmosphere during the 1950s and 60s, and 1000 times higher than that from the Hiroshima bomb. Those who received the highest doses and developed acute radiation syndromes were the people – around 400 in all – on duty during the accident, who faced the immediate emergency (plant staff, firemen and medical aid personnel) and those working on the subsequent construction of the first protection barriers. Significant doses were received later on by the large group of the so-called ‘liquidators’, who participated in several clean-up operations and carried out the construction of the shelter (the Sarcophagus). These people did not operate under emergency conditions and were submitted to radiological controls and dose limitation. Doses received by the population living around the site were lower by several orders of magnitude.

237 persons were hospitalized soon after the accident, of whom 134 were treated for acute radiation syndrome. Of these, 31 died within 90 days. Most were firemen, to whom a memorial stands today not far from the destroyed reactor. These are the direct victims of the Chernobyl accident. Other 14 people among those hospitalized for acute radiation died within ten years of the accident; not all of their causes of death can be associated with the radiation doses that they received.

The only real and, nowadays, certified effect which is likely to be related to the accident is the large and unequivocal increase in incidence of childhood thyroid cancer in Belarus and Ukraine. In these countries, 800 cases were diagnosed in the ten years after the accident, compared with a few tens of cases estimated prior to the accident. Three children had died by 2002, while generally the disease has been addressed by medical or surgical treatment. The increase in the incidence was not limited to children, as a larger number of adult cases was registered in Belarus and Ukraine, compared with previous statistics. A total of some thousands of cases is expected in those areas in the next two decades, for children who were younger than 5 in 1986.

Leukemia incidence in the contaminated and non-contaminated territories of the three states did not change in a decade after the accident, nor has there been any increase of congenital abnormalities or any other radiation induced disease in the general population of the contaminated areas, or in western Europe, that can be attributed to the exposure, seventeen years after the accident.

Since, as it is well known, malignancies due to radiation may occur several years after exposure, the late health effects have to be considered as well, along with the possible effects of the low doses being received in areas still having low level contamination. This assessment can only be made in terms of stochastic health effects, and using risk factors for calculation. These stochastic health effects are fatal cancer, serious genetically related ill health, and possible teratogenic consequences.

With this methodology, and considering the committed collective dose and a projection out to 50 years, by using conservative risk factors (which technically means using "pessimistic" figures), the calculation gives a number of *excess fatal cancers* between 7000 and 10,000, which would be added to a normal expectation of 96 million fatal cancers in the same areas for other causes (which are mostly *of environmental origin*, as it is well known). The above figure is equivalent to an additional incidence of about 0.01 %.

For serious genetic health effects, using the same method and calculation, an additional 100 cases of genetic disorder might be added to the 50,000 cases which would be expected to occur spontaneously over 50 years (an additional 0.2%).

It is undoubtedly an arid and cheerless consideration, as it always is when one tallies deaths: the number of fatalities from Chernobyl, estimated over the next 50 years, is comparable to the number killed by traffic car accidents each year in countries like Italy or France. A number that nowhere causes social concern or alarm.

***

These are the figures, or orders of magnitude, made available nowadays by science. It is difficult, however, to convince the public at large that thirty-one people died as a result of the accident, not three thousand, and that in next decades the fatalities won't be three hundred thousand, as the antinuclear activists continue to claim. But if it is difficult for the truth on the Chernobyl's consequences to prevail, it is not only because the opponents of nuclear power try to keep the public incorrectly informed.

There have been others who benefited from the pretended truth about Chernobyl, or, more precisely, who made political capital from it. In no country was this as evident as in Italy, where the most serious and definitive political consequences resulted from the accident.

It is well known that a referendum was held in Italy in 1987, when people thought of Chernobyl not in terms of the event whose origin and effect we have described above, but as an event causing fresh vegetables to be banned and milk declared unsuitable for children. A massive majority voted against nuclear energy, while major political parties, including those who never declared to be against nuclear energy, more or less officially assisted this result. Even if, technically and legally, this wasn't a vote against nuclear energy, it was taken as such, and people intended, or believed, that they must vote *pro or con* nuclear energy. The referendum was in reality about the abrogation of a law passed years before, allowing compensation for communities accepting a nuclear power plant in their territory.

The decisions taken after the referendum showed the extent to which Chernobyl played a part in the Italian political milieu. For a while, it was envisaged that Italy should maintain a minimum nuclear electricity production, from the two units of Montalto power

LIBRARY
WAUKESHA COUNTY TECHNICAL COLLEGE
800 MAIN STREET
PEWAUKEE, WI 53072
WITHDRAWN

plant, which was at that time almost 90% completed, and from the other three power stations already in operation, thus canceling the planned new plants. The intention was to keep alive nuclear power ("nuclear survival" was the slogan), which had a rather important technical tradition in the country. We shouldn't forget that Italy, to-day a totally non-nuclear country, had a strong past in this field, since it had been a pioneer in commercial power plant construction and in the development of fuel cycle facilities. If one looks at energy statistics of the fifties and sixties, one may be surprised to find Italy ranking third, after the USA and the United Kingdom, in nuclear electricity production. A trend that definitely changed during the seventies.

"Nuclear survival" was a solution supported by those who tried to save as much as possible of nuclear technology and industry of the country. Among them, there were managers of nuclear governmental agencies and State-owned companies who actually did not made great efforts to defend nuclear energy: they had perceived the anti-nuclear climate that was spreading among politicians. Indeed, the major political parties realized quite soon that, to avoid problems, it was better to finish with nuclear energy in Italy. What were these problems?

To maintain nuclear energy, even just waving the flag, it was in some way necessary to acknowledge that this energy was substan-tially safe, and do this after and despite Chernobyl. In 1988, when this discussion was taking place, the nuclear milieu of Italy already knew what had really occurred in the Soviet nuclear reactor, so technically sound arguments to defend nuclear energy were avail-able. But to lift the curtain on the Chernobyl event, the social and cultural background that was behind the tragedy had also to be un-veiled, and not all the political parties could do so.

The Left was mainly responsible for sinking nuclear energy in It-aly, and in particular the Left whose position on the energy issue had always been credible and responsive, and which had supported nu-clear development in Italy during the Montalto story and even after Three Mile Island, namely the Communist Party. The Socialist Party instead, almost suddenly had become antinuclear, well before Cher-nobyl, hoping to capture the votes of all those who became worried about nuclear energy, thus harvesting what the campaigns by antinuclear opponents had sowed. For the Socialists it was then rather easy to take a position against.

For the Communists, there was much more to it than that. Among them, there were several technical experts working in all the Italian nuclear agencies or companies, some in high positions, so the Party was perfectly aware of how things went at Chernobyl. However, to take a pro-nuclear position they should have told their comrades, worried about environment and safety, the simple but extremely embarrassing truth: in the capitalist West, including the USA where money dominated everything and where the TMI accident had occurred, in nuclear energy priority was given to safety. In the so-called homeland of socialism, not only had priority been given to production and economic needs, but the whole system had proved to be a failure.

With the best will in the world, despite their claimed "splits" from the Soviet Union, and notwithstanding the undoubted seriousness that they always showed on energy issues, the Communists couldn't declare such a truth. This is why the Communist Party of Italy, with all the influence it had on unions and the Italian society, officially declared against nuclear energy, thus forcing the other major parties to take the same position, for fear of appearing careless of environmental protection and public health.

This is the truth about the missed defence of nuclear energy in Italy, a country where almost all nuclear experts and managers were linked to the major political parties.

In other European countries and in the USA, which did not face such political problems, the truth about Chernobyl came out just when the Soviet Union was breaking up. The euphoria of the West, who believed themselves to have won the Cold War, led to a spirit of fair play and avoidance of further kicking a system that was so pitifully and obviously failing. Nations strongly engaged with nuclear energy, like France, purposely wanting to avoid an internal debate on this issue, were wary of initiating a war of figures or a controversy on the real causes of the accident. Besides, in France large nuclear industrial and engineering companies were ready to play a big role in the massive remediation and upgrading activities being undertaken in the nuclear plants of the former Soviet Union. Countries where nuclear programmes were disputed, and where antinuclear environmentalists were in governmental coalitions, like the Netherlands or Austria, far from promoting a discussion on the real origin of Chernobyl, profited from the event to phase out nuclear energy. Sweden had already settled the matter of nuclear energy by a referendum, as anticipated above; furthermore, the country was taking a key position in providing nuclear engineering assistance and services to the Baltic area of the former Soviet Union.

A very peculiar case has been Germany. In this country there were the most violent antinuclear activists of the West, and indeed the campaign on the dreadful consequences of Chernobyl had been massive, silencing any voices trying to reason about the causes and effects of the accident. Notwithstanding the wild attack, nuclear energy wasn't banned in Germany for two reasons. First, unlike Italy, the nuclear power production in the country was too important for the nation's economy to listen to the opponents' claims. Second, the two political parties in government were fully in agreement on the energy policy and on withstanding the environmentalists' pressure (although the situation did change years later, when the so-called green-red coalition won the elections).

The reason why an open discussion did not take place in Germany was rather different. In the years after Chernobyl, politics in the country was dominated by the prospects of reunification, crucial for the nation, which the dramatic crisis of Soviet Union was making more feasible each day. In those years, the Germany of chancellor Kohl was the most collaborative, we would say almost sympathetic and patient, country with the dying Soviet Union, and this wasn't a suitable climate to draw a historical and political balance from what had happened at Chernobyl. However, an idea of the perception Germany had of the Soviet nuclear technology is given by the decision taken on this matter in the country after reunification. The gigantic nuclear power station of Greinfswald, in the former East Germany, equipped with 8 boiling water reactors (not the Chernobyl type), was shut down soon after reunification, with a huge waste of money, for no reason – they were reliable reactors - but of Soviet design.

When we say that in many countries public discussion on Chernobyl's causes and effects was avoided, we mean that it was avoided just by the technical nuclear experts, since they alone were able to counter the misinformation campaign of the antinuclear groups and then to defend nuclear energy.

It is true that some decisions taken by the top management of nuclear organizations were influenced, if not dictated, by politics or commercial interests (except in Italy, where the decisions of politicians have been in some way anticipated or assisted, so to speak). Nevertheless, the disconcertment of scientists and engineers was too great to allow them to react in the short term, and many thought that, under the circumstances, trying to explain and reason was losing time. Besides, to defend nuclear energy, they should have had enter the arena and contend with the unrestrained arguments of antinuclear activists, and it is a really hard task to talk with them about *safety culture* and how it should be implemented, or about risk factors, or statistical correlations.

Therefore, the technical experts kept silent, at least in public, and devoted themselves, as we mentioned at the beginning of this book, to a kind of Lent, still continuing today, in consequence of which, since then, the antinuclear opinions alone have been reported in the official milieu, as well as in the media.

However, even if the Chernobyl accident was not brought to a real public trial, what happened during the 90s in some way helped in spreading more truthful perception of the event. The Soviet Union's failure and subsequent political troubles have been largely covered by the media, so the social, economic and environmental situation of the country in some ways has become familiar to western public opinion, along with the large environmental remediation programmes supported by the western countries. Besides, other environmental disasters have been disclosed in that country about nuclear waste management, such as discharges of high active liquid wastes to lakes.

In short, in public perception today, a severe nuclear accident in the former Soviet Union would not necessarily imply a negative opinion nor a concern about the use of nuclear energy, not unlike the way in which a plane crash involving an unreliable airline in a remote airport does not much upset frequent flyers.

Unlike nuclear technical experts, ideological environmentalists have a special sensitivity and they are rather smart fellows when dealing with communication. Slowly indeed, as the truth about Chernobyl was coming out and with it the social and economic background of the tragedy, they became aware that the event was losing much of its communication "appeal" in the struggle against nuclear energy. Then, as had happened for other reasons after Three Mile Island, the antinuclear activists have again placed the nuclear waste problem at the heart of their opposition. It is indeed the lack of a solution for final disposal that is the true, unsurmounted obstacle to the use of nuclear energy! After plutonium and the connection between military and civil uses, after the China Syndrome and the dreadful consequences of severe accidents, after radioactive clouds out of control, they ended up by discovering that waste disposal is still the most fruitful soil in which to cultivate doubts and raise social concerns.

Furthermore, they also found out that this is ground where strong opposition wouldn't be expected. As a matter of fact, fighting nuclear energy on account of waste problem does not trouble nuclear electricity production, a matter of which governments and politicians have to take care, in order to avoid complaints about energy shortages. Even if the wastes are not disposed of, especially the long lived ones which do not have a large volume (as will be seen), nuclear power plants may actually continue to run and generate electricity – it is sufficient for the time being to simply provide room for storage of spent fuel and other waste material. In those countries where nuclear energy has been phased out, the lack of a solution for wastes already produced does not have any impact on electricity production, whilst the economic problems that arise from this can only be shown by a long term assessment.

Clearly then, politicians and decision makers do not feel driven to take any action or launch programmes on waste disposal - in fact, quite often they suggest taking things easy. Yet, to avoid political trouble they have in many cases decided, as we will see, that this problem can be solved only when everyone will agree on the proposed solution.

Unfortunately, if politicians do not intend to dispose of nuclear waste in a hurry, we cannot say that scientists and engineers are showing a firm attitude to dispelling doubts and convincing people - at least in public. We might concede that such an attitude is taken by those who are closer to politicians, like the top managers of governmental agencies or state-owned companies. But the attitude of those technical experts who, on several occasions, have felt obliged to show their doubts and to appear concerned, almost not daring to declare their scientific certainties, is highly self-defeating for the social acceptance of nuclear energy. To cultivate doubt has never helped anyone to achieve consensus about anything.

Directly or indirectly, for some (not for all, as we will see), this is an attitude still to be connected to the scientific and intellectual shock provoked by Chernobyl. If, from this tragedy, ordinary people may have learnt to distrust nuclear energy, the lesson learnt by nuclear experts could be that doubt is necessary, to avoid great disappointments. However, if it is correct and moral for scientist and engineers not to take anything for granted, it must be clear that such a lesson has not been learned from that tragedy. But above all, moving from the general to the particular, we have to say that waste management, among all nuclear activities, is certainly the one leaving the least room for doubts.

As waste management is the ground that has finally been selected by the antinuclear activists for their battle - not reactor safety, always a delicate matter – then we must show clearly and without hesitation what is the state of the art in this area.

4. Technical issues

When we move through an urban environment in the developed West, particularly in one of the most congested industrial areas or in one of the more backward regions, we can perceive directly how the environment and the air that we breathe have paid the price for our life style and the goods we have available to us. No one has ever measured this effect, either direct or delayed, because this cannot be done in a reliable way. We simply know that there are widespread health effects related to what we breathe or eat, or that natural resources such as water or flora are being damaged, but it is practically impossible to establish a quantitative cause and effect relationship (that is not simply statistical), because there are too many variables to consider.

The account of deaths due to accidents connected to industrial production, including energy production, is more accessible, because it is a matter usually covered by media. Real massacres can be attributed to the exploitation of fossil fuels, particularly their extraction and transport. These losses have never entered public awareness, since they were never emphasized and never even recorded in the statistics used by environmentalists, although they can be found in those of the energy experts. To mention just one case, which is known to very few people: between 1907 and 1970 in the USA about 88,000 accidental deaths occurred in coal mines.

If negative effects on health and environment are the other face of the industrial progress, we should not fall into the trap of becoming anti-science and anti-progress. The balance is largely positive for mankind in terms of actual living conditions; indeed, the average life span, which depends entirely on these conditions, has almost doubled during the twentieth century.

But behind the animated life of a city centre, behind the activity of a factory or of a busy port, there are not only the fumes that we breath or the noises that we hear (not to mention the not infrequent offences to our aesthetic sensibility). Behind the scene there are mountains of wastes, some very toxic, arising in quantities that the average citizen can't even imagine. Moreover, what makes the volume of some of these wastes so huge (e.g. solid urban wastes), is quite often the habits of our consumer society.

The nations of the developed West have succeeded, to varying extents, to manage their wastes in order to eliminate at least the main inconveniencies and the most direct and immediate public nuisances. But no one can honestly say that a solution to this problem has been found ensuring a satisfactory safety level for both human health and the environment, especially over the long term (which is, indeed, an aspect that is almost completely ignored). This is proved by the fact that the most widely used disposal systems, such as landfill or incineration, are indeed not devoid of risks, even when managed properly. Notwithstanding, it must be noted that in political circles, in industrial production and in technological research, the safe disposal of this waste is not considered to be a real priority (even by environmentalists). Even less is the safe disposal of their wastes an essential condition for allowing the activities to continue.

Except for the nuclear energy. Here, waste has been put at the centre of the scene, more so than any other aspect of nuclear energy production, some of which are undoubtedly much more critical from a safety viewpoint. The waste issue has been so dramatized that it has entered the collective perception as one of the greatest environmental problems of our times.

Well, a simple truth must be clearly asserted. Nuclear energy not only does not produce the mountains of waste associated with other industries and with some other energy generations, but it takes care of the waste it produces in a manner that no other human activity has ever done, and probably will ever be capable of doing. And if other human activities took care of their waste as the nuclear sector does, we would all be living in a better and cleaner world.

Everybody knows this: scientists and engineers in the field, specialists from the nuclear industry, experts from safety authorities, perhaps even professional environmentalists - at least those having some scientific background. It is also known by those technical experts who, after Chernobyl, believe, as we have seen, that this certainty shouldn't be openly expressed, before a social analysis has been done, in order to allow the public to develop an awareness gradually and independently, as more or less is done with children.

It was an unforgivable weakness, for which the nuclear scientific community is to blame, not to have emphasized this fact proudly and publicly. This achievement is a result of a serious and disinterested scientific and technical effort, in which safety and environmental protection have been given absolute priority. It is equally unforgivable not to have done so in the face of the attacks of anti nuclear activists.

Even when the lack of reaction was caused by choosing understatement to avoid conflicts and make the problem less dramatic, it was a lethal choice. On one side, it gave the public the impression that nuclear experts had no arguments, on the other it encouraged opponents to continue with misinformation and nonsense. The result has been the emergence of a sort of environmental crisis around the waste problem, which is completely unwarranted.

We are convinced that the public at large, the same public that some would like to see participating in technical decisions, and others who want to carry out social investigations in order to understand how to achieve its consensus on those decisions, is capable of understanding and evaluating the reasoning of technical experts and distinguishing it from propaganda. Our society has a level of education that allows people to understand messages of this kind – provided that they are correct and clear.

***

For several years, the opponents of nuclear energy used the claimed unavailability or immaturity of technological processes for treating radioactive waste and converting it into inert material harmless to the environment as a polemical weapon. Actually, research aimed at segregating liquid and solid radioactive effluents in materials of great chemical and physical stability, potentially suitable for disposal, began in the early 60s, almost concurrently with the operation of the first commercial reactors. The United States and the

United Kingdom were the first to perform pioneering studies, later joined by France.

Since the beginning, R&D identified suitable materials for waste *conditioning* (the transformation of waste in stable solid material), which came to be used generally. These materials are cementious and glass matrixes, the first for low-level, also known as short-lived, nuclear waste, the second for high-level, or long-lived waste. The reason for selecting such different materials lay in the difference between the two types of waste, which is fundamental for the way that they have to be managed.

A nuclear waste is classified as low-level when it contains generally low concentrations of principally short-lived radioactive isotopes. Short-lived means that radioisotopes decay completely after a few centuries: short-lived radioisotopes are those with half-lives (the time they need to reduce their radiation level by half) up to thirty years. For these, their radioactivity level after three hundred years, ten times the half-life of the principal radionuclides they typically contain, decreases one thousand times (this occurs by the physical law of radioactive decay). For low-level waste, a thousand fold decrease of radiation level leads to an almost complete disappearance of radioactive isotopes; this is why a three hundred year period is the time over which nuclear waste in a low-level waste repository has to be segregated from the environment.

Instead, waste containing long-lived isotopes and/or short-lived isotopes in high concentration, is a high-level waste. For this waste, the disappearance of radioactivity or its reduction to very low levels, is achieved after a period of time going from thousands to hundreds of thousands of years. High-level waste may also have a high thermal output, due to the heat arising from the intense residual radiation (a high active liquid radioactive waste, after its production, will boil spontaneously). This is why cementious material, which may deteriorate at higher temperature, is not suitable for its conditioning. Therefore, for this waste conversion into glass is the most common approach. Glass, as it is well known, is highly resistant to heat and chemical attack (glassware has been found in ancient ships, sunk more than two thousand years ago).

Even though they have finally prevailed, cementation and vitrification have long been compared to other systems, such as incorporation in bitumen or in plastic resins for low-level waste, or calcination (i.e. thermal transformation into ceramic powder) for high-level waste.

Sometimes, even small or conjectural defects, such as unexceptional mechanical strength or suspected ageing phenomena, were sufficient to abandon materials like resins or some cement conglomerates. It is interesting to note that these materials are largely used, without restriction, for highly toxic waste, because, in this case, rigorous qualification procedures are not required.

Those who are concerned about environmental protection should know the kinds of tests and analyses a candidate material for conditioning nuclear waste has to undergo. To mention only a few of them: severe thermal cycles, long-term attack by corrosive agents, fire resistance, radiation resistance and compression resistance – even resistance to bacterial attack. It was only after technological development and investigations of this nature that the materials now being used for conditioning nuclear waste were selected.

But this is not all. The final waste package envisaged (i.e. the assembly of material incorporating waste and the outer metal container) undergoes mechanical tests using full scale models, in order to verify its integrity even after violent impacts, sinking in water, or fire, before being definitely accepted. Finally, waste can be conditioned and package by the selected process only after a public and independent control authority (at least in democratic countries) certifies and validates the quality of design and the final product.

The development of radioactive waste conditioning processes, including design and construction of commercial plants, may be considered to have been achieved by the middle 80s, when industrial scale vitrification plants began operation. These plants, prevalently using French technology, were those that required the greatest development effort.

It is several years since anyone, even the most obstinate opponent of nuclear energy, could pretend that a reliable technology for conditioning nuclear waste, including high activity waste, is not available.

In nuclear plants worldwide, including those already shut-down, large quantity of low-level waste packages have been and are being produced, along with a number of high-level waste vitrified canisters. Therefore, radioactive waste conditioning has, for a long time, supplied no more arguments to the opponents of nuclear energy.

The arguments today concern, instead, the final destination of nuclear waste, namely the transfer of waste packages to an appropriate site where they are disposed of permanently.

In a disposal system, as the radiation associated with the waste lasts for long periods of time (ranging from centuries to many millennia), an assessment of environmental effects has to be made that extends to the same periods. In order to do this, all possible physical and human-induced transformations of the site and its environment must be taken into account over that time. In such assessments, particularly when the evolution of human factors are involved, there is inevitably room for many non-technical considerations, from sociological to ethical, and it is not difficult to raise doubts and uncertainties and to let the scientific and technical basis of the concept to be shadowed. Nevertheless, science and technology have also succeeded in providing sound solutions to this problem.

***

The principle governing nuclear waste disposal is that the package must be placed in a repository in such way that the harmful materials it contains are prevented from coming into direct or indirect contact with the so-called 'biosphere', namely the living environment around us, and, through this, with people. Another basic principle is that this contact has to be prevented as long as the waste continues to be dangerous.

Isolation from the biosphere is obtained by placing several containment barriers between the harmful materials and the external environment, arranged in succession so that each barrier reinforces the preceding one, whose function is to avoid release of radioactive isotopes from the facility in any foreseeable circumstance, both normal or accidental. Considering that the only common natural agent able to mobilize and transport radioactive material is water, by a dissolution process or merely by physical entrainment, the actual function of the barriers is to prevent or minimize water flow in contact with the radionuclides of the conditioned waste, or anyhow to avoid any release from a repository should this contact take place, including as a result of accidental events.

The first of the barriers is the waste package itself, produced by the conditioning process, whose material is selected, as we have seen, just for ensuring the segregation and immobilization of the radioactive material. Other barriers are provided by the repository structure and they can be either man-made or natural, or a combination of both. The repository safety, both in the short and the long term, is based on the reliability of these additional barriers, whose nature depends on how strict the containment needs to be, and for how long it has to be effective.

The main effort in devising, designing and constructing a repository for low as well as for high-level nuclear waste is a technical activity that goes under the name of *qualification* of these barriers.

What is the barrier qualification? It is the whole activity dealing with the selection of barrier materials, the study of their properties, including evaluation of their long term behaviour and experimental tests carried out to confirm them. All these actions are aimed at ensuring that barriers will actually be capable of maintaining the waste completely isolated during the required period and limiting any releases thereafter.

This is where a substantial difference between short-lived and long-lived waste occurs. The radiation of the first will disappear in a few centuries; the latter, as previously explained, loses its radioactivity only over a very long time, tens of thousands of years or more; it so needs to be isolated for a period of time of the same magnitude. In the two cases, the performance of the barriers has to be quite different.

For low-level waste, about 95% of the whole production of nuclear wastes, isolation must be assured at most for a few centuries (three hundred years is the usual reference time, bringing on a thousand fold reduction of radiation level in the principle radionuclides, as we know). This is a period of time during which it is certainly possible to ensure conservation and duration of engineered barriers, provided they are adequately designed and constructed. Concrete is used for this type of barrier, either reinforced or not. Its mechanical, hydraulic and chemical properties make it an ideal material to guarantee defence not only against massive water infiltration but also against mechanical impacts or penetration.

The longevity of such properties – in this case for at least three hundred years – is obviously the main requirement that both the material and the whole structure have to meet. For this reason, not just any concrete is used, but a special mix in which cement, inert fillers, chemical additives and the steel bars, in case of reinforced structures, are carefully selected and tested. Estimation of their long-term behaviour is made by so-called accelerated tests, whereby the material undergoes extreme tests directed to simulate the effect of stress that, in reality, are spread over considerable periods of time (for instance, immersion of specimens in water for months simulates the effect of century-long infiltration of rain water or humidity from soil).

Three hundred years is not a prohibitive nor a critical duration over which to assure the efficiency of materials and structures made from concrete. Today, lifetimes of centuries may be required for certain buildings or civil structures. A recent example is the dome of the new Great Court of the British Museum in London, which had a design requirement for a guaranteed lifetime of two hundred and fifty years. The most impressive evidence of the durability of concrete structures remains the dome of the Pantheon in Rome, built at the time of Hadrian, about 150 A.D, and never subjected to significant restorations. The dome was cast using a rather rough grout, by no means comparable to that produced with modern cement.

The reliability and safety of short lived nuclear waste isolation from the external environment, for the whole time necessary to decay, can thus be achieved without difficulties, neither of material selection nor of engineered structure design and construction.

Actually, disposal of this waste has been or is being carried out in many countries, in practically all those using or having significantly used nuclear energy, with the exception of Italy. The repository structures in most cases have been built on or in proximity to the surface (they are called *near surface* repositories, or *vault* repositories, and can be built below or above ground). In Scandinavia, there are also underground repositories, built with the same criteria (i.e. isolation is principally by man-made barriers). The Swedish repository, positioned under the Baltic Sea, about one kilometer from the coast, is a spectacular one. A case apart is Germany, where a worked out iron ore mine is used on account of its extremely favourable mechanical and hydrogeological features.

In its more general configuration, a near surface repository is a succession of reinforced concrete cells in which waste packages are placed. The empty space between cell walls (forming the outer barrier) and packages is backfilled with waterproof material, acting as further barrier. The whole structure makes a multiple barrier system (the classic one, described above, has three barriers), where each one by itself would be capable of providing effective isolation of radionuclides from the environment. The arrangement in series ensures that containment remains should the previous barrier fail or weaken. In such system, the repository cells, once filled with waste packages and with backfilling material, become concrete blocks of great size and stability, where an overall barrier from one to several metres thick is interposed between the radioactive waste and the external environment. Radiological safety, namely the efficiency of barrier and isolation, is continuously controlled by an environmental monitoring network, including the repository as well as the surrounding area. Furthermore, these repositories are located at sites having favourable geographic features, able to provide an additional defence (such as an impermeable soil beneath the repository, preventing any migration of radioisotopes).

Disposal facilities of this kind are in operation or under design in all countries where low activity nuclear waste has been produced. The most modern are in France, Spain, Sweden, Finland, Japan, the United Kingdom and the USA. Important projects are also at an advanced stage in Germany and in some Eastern European countries.

In addition to the actual disposal units, the repository usually comprises other facilities such as waste conditioning stations, analysis and control laboratories, remote-control transport and handling systems, utilities and administration buildings, sometimes also a research laboratory and quite often even a visitors centre and guest house. The whole area actually comes to be a kind of Hi Tech Centre, where qualified technical and scientific activities are performed.

These are the ill-famed *nuclear dumps*, as are frequently highlighted in newspapers and the media, which prefer the vocabulary of the antinuclear activists over that of engineering.

We can say, without fear of being challenged, that all radiological controls and environmental monitoring around such repositories are aimed more at reassuring the public than at anything else. External radiological effects due to disposed waste are indeed non-existent; on the other hand, it is hard to imagine how concrete blocks, where radioactive material is confined within walls several metres thick, might have an environmental impact; not to speak of layers of earth placed on top of the repository, aimed at restoring the previous landscape.

Those who have some familiarity with real waste dumps (not to mention those obliged to live near them), or simply aware of the problems related to conventional waste disposal, could now understand, provided they are intellectually honest and have some technical education, why we began this explanation by declaring that no human activity takes care of its waste as nuclear energy does.

We can easily say, again with no fear of contradiction, and we might even demonstrate it by calculation, that a petrol station in a city releases in *one day* to the public, and mainly to the staff, as much poison as a near-surface repository for low-level radioactive waste will never do during the three hundred years over which it isolates waste.

The principle of isolating radioactive waste in a repository by barriers, either natural or man-made, also known as the defence in depth principle, is not simply a feature of the repository design. It must also undergo validation based on mathematical criteria, which means that it has to be shown quantitatively that there will be no harmful release for man or the environment from the repository, in any situation, even the most critical. This is an essential condition for having the repository design approved.

We will try to explain how this environmental impact calculation is made and the criteria applied. It is indeed one of most advanced calculation methods one can imagine and, once more, never attempted for any other industrial activity, including the most hazardous ones.

Let's consider, in a near-surface repository, a radioactive element (say Caesium-137, one of the best known isotopes, whose radioactive emission is reduced by 50% every thirty years) placed at the centre of the repository cell, isolated from the external environment by concrete layers. If not reached by water coming from the outside, it will never move from its position, except for diffusion through concrete pores containing water that is not free, but chemically bonded (the water needed to hydrate the cement and to allow hardening). This mass transfer process can be calculated by the laws of thermodynamics and it amounts to a few millimetres over thousands of years. Nevertheless, concrete can deteriorate (very slowly indeed) by various processes: for instance by carbon dioxide absorption from the air or ground or other chemical agents, like sulphur dioxide and nitrogen oxides, and this may favour the migration of some chemical species which, under normal conditions, would not move. The rate of this absorption, and consequently the level of possible degradation, is known from the scientific literature and from experimental tests carried out to select the most appropriate grout. Even in the most severe conditions imagined, a radionuclide like Caesium can travel only a few millimetres before it decays to harmless levels. Caesium, if there is no water, can never escape from the concrete barriers.

But the case of no water entering from the outside is a rather academic case, considered just because in nuclear energy everything has to be taken in to account. But how and why could water reach Caesium-137, which is confined inside the repository cell and protected by various layers of cementious materials? Actually, water *cannot* reach Caesium, under normal condition, for three reasons: the repository is built on a site that cannot be flooded, because of its geomorphology; the repository is built on a site whose superficial and underground hydrogeology makes contact with underlying groundwater impossible; the waste package is protected from rain water by an engineered cover made from several layers of cementious material, which is a *hydraulic* binder, i.e. it acquires its waterproof properties just by contact with water. While the first two conditions do not change with time, unless important modifications in the environment take place, the third is subject to change, because, over long periods of time, the artificial barriers may deteriorate. In addition, they may lose their containment efficiency, partially or totally, because of accidental events.

The events that, on paper, may cause a failure of engineered barriers and therefore expose waste to contact with external water, are identified and characterized by a procedure which, in the design of nuclear waste repositories, is known as *definition of scenarios*.

Scenarios are those normal and extraordinary situations, evolutions and events a repository may have to face during its entire lifetime. A *normal* scenario occurs when, over the whole period – three hundred years - the repository evolves in a natural manner, in other words it operates according to design specification and undergoes the expected degradation rate for cementious structures (due to a natural aging process). If, for instance, a loss of impermeability of 10% every hundred years is selected as a design basis (a degradation, anyhow, uncommon for this material), then calculations must prove that even at this degradation rate there is no release to the environment. But this is not sufficient. To make the scenario more severe, the worst situations that may accompany the barrier degradation are also envisaged. So, in evaluating water infiltration and its effect on the solubility of waste material, the heaviest rainfall is assumed, or the most harmful radioisotope considered affected, or the drainage system installed below and around the repository, to collect and treat infiltrated water, is considered not to be working. In every situation of a given scenario, it must be proved either that nothing is released from the planned repository or, should migration of a radionuclide from the repository take place, that there will be no radiological effect on man and the environment.

It is worth noting that calculations made for multiple barrier repositories being built today, show that, even assuming the most unfavourable chain of events and the most severe degradation with time of structures and barriers, the effect on the environment and health is minute. The radiological doses resulting from these calculations, for a normal evolution scenario of a near surface repository, are very much lower than those due to natural radioactivity or cosmic rays.

The same approach is applied in evaluating the behaviour of the same Caesium-137 in the case of "perturbed" evolution scenario, whereby abnormal or accidental situations are assumed taking place in the repository. The most severe event considered leads, for whatever reason, to sudden and complete failure of the barriers and consequent exposure of radioactive waste to atmospheric agents. It should be noted, first of all, that a low-level waste repository is made, when filled and sealed, of a number of very large and stable concrete blocks, with an earthen cover. High temperatures and pressures, as found in chemical or nuclear power plants, are not associated with such systems. Any traumatic event, such as a mechanical impact (like an airplane crash) or an explosion, cannot have dramatic effects. However, any event envisaged for perturbed scenarios must be described and discussed, in order to prove that the planned defences are able to withstand natural or human induced events, preventing the release of radioactivity to the environment or, in any case, avoiding radiological doses to the population.

Once more, we want to stress a concept: it would be desirable, for the benefit of the environment, to see the same care, taken for Caesium-137, being applied to many metals contained in conventional waste, such as lead and mercury, to mention only a couple. Though they are certainly not less harmful to man's health than Caesium, materials containing such metals are disposed of in conventional landfills where, even in the most advanced ones, no precaution is taken for their long term destiny. And these metals, unlike radioactive Caesium, will still be around after three hundred years, even if no one will know where and in what condition. What is certain is that nature will not have helped in providing their disappearance, as is the case for Caesium-137.

We elaborated on the subject of a low-level radioactive waste repository because the construction of such a facility – and therefore the selection of a suitable site - is a chapter still to be written in Italy, where inevitably we will see enraged environmentalists at work, doing their best to spread fear and dread among the population involved, and turning their understandable concern into opposition to a repository's location. And we will also see, very likely, the disengagement of local and national politicians, should emotion prevail and the opposition become too noisy.

Nevertheless, to conclude this technical discussion, a truth has to be stated once more, one that is supported by scientific evidence: there isn't in Italy, or elsewhere, even in the most developed areas, a local administration or community having within its territory environmental problems that are less worrying than those due to a possible radioactive waste repository. It is sufficient to consider the pollution due to the traffic in the main street during the rush hour.

***

In the final repository for long-lived waste, as mentioned above, waste must also be isolated from the biosphere for the whole period it is still harmful for man and the environment. It is indeed a much longer period of time: thousands, tens or sometimes hundreds of thousands of years, since this is the number of years over which long lived waste continues to have a dangerous radiation level. As it will be better seen in the next chapter, on account of this time scale, in recent years a sociology has flourished around long lived disposal, a very peculiar sociology indeed, thanks to which remarkable scientific and technological achievements have been pushed into the background. And yet, technical solutions have been extensively studied and carefully tested for the disposal of long-lived waste too; solutions that, notwithstanding some uncertainty - unavoidable under the circumstances, as we will see - can be considered as the most advanced that science and technology could have conceived for the safety of present and future generations.

Here again, the criteria of barriers interposed between the material to be isolated and the environment is applied, and, in this case too, the barriers are aimed at avoiding contact between waste packages and water, which could carry radioisotopes into contact with the animal and plant world.

Can man-made barriers be efficient in preventing water infiltration for tens or hundred thousands of years? The answer is not necessarily negative, even though, as we will see, sole reliance on engineered structures is avoided. It is well known, for instance, that there are metals practically unalterable with time, like copper and some titanium alloys. Materials of this type are indeed planned for manufacturing special overpacks in which waste packages might be placed, in order to supply a further barrier around the waste package itself.

But before going further, we should discuss the nature of the package to be disposed of in the case of long lived waste.

As discussed in the first chapter about the nuclear fuel cycle, as reprocessing has been abandoned in many countries and is being questioned as an industrial practice in others, spent fuel is becoming the predominant high level material to be disposed of, while glass canisters (used for high-level waste from reprocessing) will continue to be produced, at least in the medium term, in France, the United Kingdom and Japan. Particularly in the USA, where reprocessing has been abandoned since 1976, a geological repository for spent fuel is under licensing procedure, as we will see later in some detail, while in Sweden and Finland a similar disposal system is being planned.

For disposal of high level waste then – either radioactive waste conditioned in glass matrixes or spent fuel unloaded directly from power plants - barriers must be provided to ensure isolation from the external environment, which means preventing water infiltration, for tens of thousands of years and more.

The solution pursued nowadays to achieve this objective is disposal in deep geological formations which, by their nature and origin, are considered generally suitable because they are expected to remain stable and unaltered for periods of time measured on the geological time scale, such as hundreds of thousands, and even millions of years.

These requirements are met by some sedimentary bedrocks, in particular salt formations, formed during the long evaporation process of former oceans, and by clay basins. Also suitable are particular types of crystalline rocks, such as non-fractured granites. Salt formations are an ideal medium: their very existence testifies that water has kept away from them for geological ages, as salt (in most cases, the same sodium chloride used for food) is highly soluble in water; its evaporation was in fact the process that caused the bedrock's formation. Besides, the salt rock is rather plastic, therefore capable of absorbing possible tectonic stresses at its boundaries without any mechanical damage. Clay formations have equally favourable characteristics; according to some experts, even better ones (as is natural, both types of bedrock have fans). On the other hand, it is true that clay, besides its well known impermeability, behaves as a geochemical barrier, which means it can chemically absorb migrating chemical species.

Like for a low level waste repository, the performance of a geological repository's barriers must not only be defined carefully, it must also be proved. In order to do this, materials – salt, clay, granite and others - have first been extensively tested in research laboratories, a normal procedure for materials science; then, underground laboratories, hundreds of meters deep, have been excavated directly in the bedrock formations to be investigated. Experimental activity carried out in these laboratories, extending from hundreds of metres to some kilometres, is aimed at evaluating the behaviour of the natural rock barrier when in contact with the material to be disposed of, which may have a non-negligible thermal content, and at verifying *in situ* the bedrock's physical and chemical properties. The rock engineering problems connected with the construction of a repository are also assessed thanks to the underground facilities.

Experimental mines and laboratories of this type have been built since the 70s in Belgium, Germany, Sweden and Switzerland, and one is under construction in France. In Germany a large salt mine, one of the best (geologically) worldwide, has been excavated and operated for many years as an experimental facility.

So much information has been gathered on geological formations and rocks that, today, constructing a deep underground repository -

between five hundred and one thousand meters below ground - presents no particular uncertainties, neither in design and engineering, nor in the chemical, physical and geological qualification of the barriers.

If stable and impermeable rocks are the ultimate barrier to rely on for permanent isolation of waste, man-made barriers are not neglected. An additional metal container is planned, aimed at reinforcing the first barrier (the steel canister containing the glass in the case of high active waste, or spent fuel in its metallic structure) and making it more resistant to the selected geological environment. In some cases, the outer container is also designed to help making the disposed package retrievable. More robust barriers and provisions for waste recovery are requirements, as will be discussed later, which come from non-technical issues.

Some metallic materials under consideration as outer containers (overpacks) in the engineered barrier systems for long lived waste (titanium alloys, special steels, copper) would be able, by themselves, of ensuring isolation for many thousands, even tens of thousands of years, when of a suitable thickness. A special metallic overpack envisaged in Sweden and Finland for the geological disposal of spent fuel, made of copper and manufactured using a special technique with no welding (welding favours corrosion attack) can by itself provide isolation for a million years, as calculations based on corrosion kinetics show! Sweden and Finland are, however, a particular case: the geological formation considered is granite, widespread in Scandinavia, but it is a rather fractured bedrock, hence water infiltration cannot be excluded in the long period. Engineered barriers are then planned, that include naturally occurring material such as bentonite, aimed at reinforcing the isolating capacity provided by geology.

For geological disposal, as we have seen for a near surface repository, safety and reliability of containment must be proved by a calculation methodology directed to assess the barrier performance in all foreseeable situations, either normal or accidental, and to evaluate the radiological impact of possible radioisotope releases.

The definition of *scenarios* for the evolution of a geological repository is an as yet discussed matter, mainly on account of the time periods involved. This is an area where, as previously mentioned, sociological exercises of various kinds are possible (often assisted, as we will see, by technical extravagances). We will examine these positions, and their actual and possible consequences not only on radioactive waste disposal, but generally on nuclear energy, in the next chapter. Here we want to concentrate just on technical issues.

To define evolution scenarios means, first of all, establishing to which period of time the repository's performance assessment must be extended.

As mentioned above, waste isolation from the biosphere must be guaranteed so that no radiologically harmful consequences ever occur. As an example, let us take the case of two long-lived isotopes, plutonium-239 and carbon-14, both contained in this waste. The first has a half-life (the time necessary for its radioactivity to be reduced by one half) of 24,000 years, the second of about 5,700 years. For them, the time required to reduce their activity to one thousandth is 240,000 years for plutonium and 57,000 for carbon-14, namely ten times their half lives.

It is clear that, even though it is easy to show that one of those geological formations described above will remain unaltered for such time periods, and therefore will ensure isolation of plutonium and of carbon-14, it is quite hard to prove it by calculation.

Why? Because the earth sciences, involved in these evaluations, provide scientific instruments of a deductive nature, suitable to describe the evolution of our planet's past. When evaluating the evolution of a repository in a quantitative manner, namely using numbers and not qualitative estimates, an inductive method has to be applied, which, by definition, implies an inevitable degree of uncertainty. In other words: we can certainly say, and it would be true, that a deep salt or a clay formation, wherein a nuclear waste repository is located, will ensure isolation from the biosphere for hundreds of thousands of years or more. We may say so because the geological history of the earth tells us how bedrock is formed, and from the same history the expected future evolution of the bedrock under consideration can also be known. But, as we now know, with radioactive waste and with nuclear energy in general, expectations, though scientifically founded, are not considered sufficient. It is necessary, at least, to establish *how much* they are scientifically founded.

Nuclear energy has not refused this challenge. A special scientific activity was undertaken a couple of decades ago, directed at tackling the uncertainties existing in long-term evaluations of long-lived waste disposal. By using experimental facilities as well as mathematical models, a geological repository system - the structure conceived by the mining engineering and the surrounding natural environment - has been studied in all its aspects. From thermodynamics to hydrogeology, from radioisotopes migration and absorption to heat transfer, all the properties of geological formations and their evolution with time have been assessed and are still being evaluated by the most advanced techniques, mobilizing all the scientific disciplines. The experimental work under way in the underground laboratories above mentioned is also directed at validating the calculation methodology applied for this assessment, besides appraising, on site, the behaviour of materials and structures.

With regard to the evolution scenarios of the repository, all the possible critical situations, wherein barrier failure could occur, have been considered. Are there events that can modify the geological medium, perhaps unexpectedly, and, if so, what is the probability of their occurring? How many thousands of years does it take plutonium-239 or carbon-14 to move in salt or clay, supposing their container is destroyed completely? Can the effects of long-term climate change such as glaciations or desertification, modify the behaviour of deep geological formations? What happens if an inadvertent intrusion, for instance oil drilling or mine excavation, takes place when historical records of the repository have been lost?

The answers to these questions, and to others of the same kind, have to be given, and indeed can now be given, in order to evaluate the long-term safety of a geological repository for nuclear waste. And for each event considered it must be proved that a possible failure of one or more barriers does not cause radioisotope release, or if a release should occur, that there will be no harmful effect on health and the environment. And this must be shown to be valid, as we have seen for a near surface repository, for the whole period that waste has to be isolated from biosphere.

How long this period must be, is an issue long debated, and still being debated. As a matter of fact, this is a subject where some have gone too far, entering a kind of anthropological and scientific Byzantinism, pretending that it has to be ensured, which means *proved*, that a geological repository will remain stable for millions of years. This is a point where different schools and positions exist, in particular between Europe and Japan on one side and the USA on the other; positions reflecting a different cultural approach to the problem rather than a different position towards safety and radioprotection. In the USA, the assessment of geological repository performance, and therefore the efficiency of the provided defences, goes only as far as ten thousand years, and no account is taken of possible waste harmfulness after that period. This approach to the time scale problem is based on the consideration that projections over ten thousand years, while possible with the instruments mentioned above, make no sense and have no scientific validity. They believe, not wrongly indeed, that too many unforeseeable changes may take place in society and in the habitat during that period of time, to make the results of evaluation methodologies, that *anyhow refer to today's life styles and social and cultural settings*, significant.

An example may help clarify how assessments are sometimes performed which, on close inspection, appear to be completely senseless. In order to calculate if a possible return to the biosphere of a radioisotope can be harmful for health, the criterion of the number of fatal cancers induced by radiation is applied. The limit to be met is one case of cancer per one hundred thousand inhabitants (in some cases one per million). Can this criterion have any sense when applied to an event that may happen ten thousand years from now? Not only is any prediction on the social life in that era arbitrary (even assuming that there will be any, as such), but to apply the above criterion on fatal cancer numbers, it means considering this disease as having, ten or fifty thousand years from now, the same mortality rate it has today, which is a senseless estimate.

Anyone can easily understand that not everything can be predicted and evaluated, when dealing with such a remote future. And yet, this is what is continuously required, and this lack of certainty is quite often invoked as a pretext for doing nothing or for postponing decisions.

Never in the history of mankind has a need for safety been stipulated that is as stringent as that required for the disposal of the few thousands of cubic meters of high-level waste being produced worldwide by nuclear energy. Nothing like it has ever been asked, not even for events or situations that are far more important and decisive for the common good and for the future of mankind and Earth - events and situations whose control and mitigation might perhaps justify a similar concentration of efforts and resources and such severe determination as that required for these unfortunate long-lived wastes. Just to mention a few of them, very popular with the daily news: water resource depletion, earthquakes, especially those in densely populated urban areas, endemic and epidemic diseases, floods. And we did refrain from mentioning some others, still controversial but already beloved by ecological catastrophe preachers, such as the greenhouse effect and long term climate change.

We may conclude this chapter, devoted to technical and scientific issues, by reasserting that, for long-lived nuclear waste, as well as for the short-lived, science and technology have provided safe solutions, and that the remaining areas of uncertainty are, in any case, dealing with situations and circumstances which it is impossible to scope reasonably, because it would not make sense. And, finally, let's recall that there are, in the world we live, much greater uncertainties with which we are used to living without problem – in fact we almost ignore them.

On technical grounds therefore, not unlike the case of the large bridge we mentioned at the beginning of this book, nobody doubts that scientific and engineering knowledge is available to start construction of safe geological repositories. Some countries have already decided to go ahead, as we will see, and have begun site characterization and design activity. Others could do so, because they have the capability, but they are waiting for better times, which means they wait for more favourable conditions for a repository's public acceptability. Actually, for them the challenge is just a political one.

5. Nuclear Waste and Democracy

Nuclear waste is not only a by-product of industrial and experimental operations – a by-product that science and technology take care of in a way that has never been even attempted for other human activities. Nuclear wastes, in particular those requiring a geological repository, are also materials the disposal of which often requires the application of a new concept of democracy.

In recent years, a school of thought has arisen and spread, somehow cultivated by nuclear experts and managers rather than by environmentalists, claiming that science and technology's achievements are not a sufficient basis for nuclear waste geological disposal. Particularly on account of the very long time periods involved, *society as a whole* must become confident in geological disposal, and what has been achieved by science and technology has to be shared and accepted by the public before going ahead.

Whilst for some this attitude simply reflects the practical and rather understandable problem of gaining consensus on decisions that could be very controversial, others *purposely* emphasize the uncertainties affecting the long term projections that we have mentioned in the previous chapter.

Perhaps the most numerous group takes this attitude simply to procrastinate and postpone decisions that might raise opposition, or make them less disputed. Even while obeying a political logic, they do not disclaim the scientific achievements. There is no doubt that, without any hesitation whatsoever, they would move to site selection and construction of a geological repository, could such a decision be taken without creating discontent and political difficulties. The scientific knowledge and engineering capability certainly exist to do this. Several countries are in such a situation, the most evident cases being France, the United Kingdom and Sweden.

The others, those inclined to emphasize the technical uncertainties, have rather more oblique motivations and objectives. While it cannot be excluded that some are still suffering from the consequences of the Chernobyl tragedy, and still under the influence of that "ethic of doubt" that has touched many nuclear experts since then, the reason most probably lies elsewhere. When a position of this kind is taken by representatives from governmental agencies, it just reflects the antinuclear policy, or simply a suspicious attitude towards nuclear energy, pursued nationally by the governing political parties. This is the case in Germany in recent years, where the Greens have been in government coalition. Instead, when these doubtful attitudes find citizenship in academic circles or in international agencies, they often reflect a form of scientific narcissism, which ends up by becoming a sort of byzantinism. Both in history and science, this has always been inspired by more or less conscious needs of self-conservation or self-promotion.

Whatever their intention may be, the usefulness of these positions to the opponents of nuclear energy, is easy to recognize. To argue that, for geological disposal, science and technology achievements are not sufficient, endorses the view, now accepted by some in nuclear circles, that considers the problem of radioactive waste disposal unsolved, and at this point almost insoluble.

This is indeed the frontier, as we tried to explain above, where the anti-nuclear opposition has definitively established itself (even forgetting Chernobyl), on account of its potential promise. That we cannot rely on science's achievements (for waste disposal as well as other issues in nuclear energy) is a statement which has been the preferred weapon of the ideological opponents of nuclear energy since the beginning. In fact, it is on this ground that in some countries there is a policy or a call to abandon nuclear energy and sometimes even to shut-down operating nuclear power plants.

There is another aspect, not to be disregarded, which may have more practical consequences. This attitude towards geological disposal, that lies somewhere between doubting and waiting, may also influence policy and programmes on short lived nuclear waste disposal, where this problem is yet unsolved, such as Italy.

Whatever the background, technical or political, of the request to involve the public in the decision making process, the conclusion for this school of thought is just one: decisions on geological repository construction, and in particular that mother of all decisions - *where* -

may only be made when there is *societal* acceptance about them. In waiting for this, a temporary solution has to be found.

All this, despite the years, many years, spent studying all the scientific and technical aspects of geological disposal; despite the impressive underground laboratories excavated to test and analyze everything, and the experimental geological repositories operated in the past; despite one presently in operation (in USA, as we will see later); and despite the sophisticated methodology developed for safety assessment.

For long-lived nuclear waste, this is not enough. It is likewise insufficient to discuss and approve an identified solution according to the usual procedures of representative democracy. This is not, in fact, what this philosophy means by public participation and involvement. Instead, a virtuous route has to be taken which, implementing the new concept of democracy mentioned above, would allow people's confidence in the proposed technical solutions to be built up, so that geological disposal would not only be scientifically safe, technically feasible, and approved by the people's representatives, but also accepted by so-called civil society. Only at this point can a geological repository be realized.

***

The most eminent place where nuclear waste geological disposal has progressively lost its scientific and technological traits and become a societal problem, is the Radioactive Waste Management Committee (RWMC) of the Nuclear Energy Agency (NEA) of the Organization for Economic Cooperation and Development (OECD). This Committee may be considered a sort of scientific body where, for many years, the problems of nuclear waste disposal, in particular long-lived waste, have been debated. In the NEA, all the main western nuclear countries and Japan are represented, including those - as is Italy's case - having played a role in the field historically.

It is within the NEA, and particularly in the RWMC, that the principle has been cultivated and has found citizenship, according to which scientific certainties and achievements do not suffice for nuclear waste geological disposal, and where new frontiers of democracy are being invoked, claiming the involvement of the public and civil society in technical decisions.

In this committee, this procrastinating attitude can be quite easily recorded among the representatives of those countries which, although they have conspicuous quantities of long-lived waste and important geological disposal programmes under way, for one reason or another are not inclined to making a decision. It is also the case here that if some country, used to practicing democracy in a more classical way, decides to take action on a geological repository, it ends up by being accused, more or less covertly, of taking a dangerous short-cut.

This is what happens with the USA, where, during 2002, crucial decisions have been made about geological disposal. Following years of evaluations, researches, several legislative measures and not a few disputes, the Department of Energy (DOE) decided, according to a procedure established by law, to recommend to the President a site in the desert of Nevada, called Yucca Mountain, for locating a geological repository. Two decades after having been appointed to this task, under the well-known Nuclear Waste Policy Act, the DOE has finally selected the site.

Those who are familiar with the evolution of the US project, know that this step was preceded by a long and thorough debate, characterized by several public hearings, both in Nevada and nationwide, in which independent scientific institutions have been also involved – among them the National Academy of Sciences. As provided for by the law, the President can either reject the proposal, or accept it and himself recommend the site to Congress, for it to decide and approve the selection via a congressional Bill.

George Bush, who received the DOE's recommendation on 15 February, sent it to Congress the day after, with a readiness wholly unusual even for routine measures and this certainly was not one of those. The Governor of the State of Nevada vetoed the choice, as the Nuclear Waste Policy Act allowed him to do within 60 days from the President's recommendation. But the same law also envisaged that the disapproval of Nevada could be overruled by a majority vote of Congress and Senate within 90 days of continuous session. This is what occurred in July 2002, when the veto was rejected by 306 votes against 117 in Congress and 60 against 39 in Senate. Since then, the Yucca Mountain Repository project has begun the long licensing procedure aimed at obtaining the construction authorization from the competent federal authority for nuclear safety. This procedure, it is worth recalling, may also lead to the project being rejected, but this would only be based on technical issues, not on political, psychological or social ones.

This procedure of very ordinary democracy is perhaps the furthest imaginable from that consensual and participatory route, involving society as a whole, that is recommended by that sort of nuclear Byzantium which has become the Radioactive Waste Management Committee of NEA. Indeed, beyond the official reactions, which were in any case not at all enthusiastic, the US decisions have been welcomed by NEA's RWMC with some grumbling, because the *people* would have not been sufficiently involved in the decision making process. ("Gentlemen, for us the people is the Congress" – was the answer given to such comments by one of the American representatives.)

Yet, in the USA, the matter of long lived waste disposal had been under discussion for years, as everybody knows, involving all organizations interested to the problem, including the most radical anti-nuclear activists, such as Greenpeace and the Sierra Club. But then the moment came for decision. Thus Americans, perhaps more similar to ancient Romans than to ancient Byzantines, can hardly imagine that a more democratic decision making procedure has to be envisaged to construct a nuclear waste geological repository than the one they apply to make war.

So, for long-lived nuclear waste, science's achievements are not enough. It is even insufficient that a selected solution be discussed and approved by the highest political assembly of a nation, the Parliament.

Let's see what is really needed to make decisions about nuclear waste geological disposal, according to a well known NEA document published in1999:

> A general agreement and confidence in society about the ethical, economic and political aspects of the geological disposal solution [...]
>
> A wide confidence basis on the practicability and long term safety of geological disposal [...]
>
> A wide confidence in organizational structures, on the legal framework and on established licensing and review procedures [..]
>
> A careful, incremental approach to decision making, with the possibility of reversing decisions at any moment [..]

This is not all. Let's see once more what – as stated in an NEA workshop held in 2000 – the behaviour and attitude of managers, experts and organizations appointed to run these long term projects should be:

> Organizational features to include a structural learning capacity and an internal culture of "scepticism", allowing practices and beliefs to be periodically reviewed, strong internal relations and cohesion, consciousness of the ethical dimensions of their role [..]
>
> Behavioural features to include openness, transparency, honesty, willingness to be "stretched", freedom from arrogance, recognition of limits, active search for dialogue, an alert listening stance and caring attitude [...]

We may stop here, even though the NEA's literature on such subjects is rather lavish.

We should wonder, when confronted by positions of this sort, how they could possibly have been devised by technical experts (since it is of technical experts that we are speaking), who, for the generation to which they belong, the countries from which they come (the NEA only comprises democratic nations), have done all but participating in non transparent or socially dangerous activities. Because only those who are accustomed to experiences of this kind, or who are used to working in authoritarian societies, can be requested to follow the types of rules, procedures or simply recommendations mentioned above. What is most intriguing, forgetting for a while the damage they can provoke, is the mention of the "culture of scepticism", which, to ordinary people like us, recalls the philosophy of the great Greek sophists, one of Western culture's great sources, actually having nothing to do with feeling bound to doubt scientific achievements.

Further on we shall see what, in our opinion, is the explanation of this *dérapage*, as the French say, of these technical experts. Here we want to give a picture of where this attitude, which nowadays we might call fundamentalist, has driven nuclear waste geological disposal in recent years.

As was inevitable, the Decalogue mentioned above not only involves a behavioural message (such as scientific scepticism, and those virtues we may synthesize as goodness of heart), but there are also momentous practical outcomes.

It is evident that, for the above mentioned conditions to mature, especially those related to the public's confidence in technical solutions, not only is time needed, but the technical solutions themselves have to help people, so to speak, to become confident. Thus, what originally was the disposal of nuclear waste (properly conditioned) in a suitable geological bedrock, adequately prepared by mining engineering, has become, in many cases, a repository where the geological barriers are no longer sufficient, but other, so-called more robust barriers, are required. This means that, in the hope of making the geological repository, certainly not safer, but more acceptable, additional man-made barriers are demanded. More than reinforcing waste isolation, these are avowedly aimed at reinforcing public confidence. But this is not all.

After more robust barriers, came retrievability – the possibility of retrieving waste after it has been disposed of.

Everyone knows that, from a safety viewpoint, this is neither necessary, because the geological barriers will assure a more unalterable isolation in time, nor desirable, because returning to manipulate radioactive substances, after a long time period, may be rather complicated and for sure not devoid of radiological effects. Retrievability too is directed just to soothe people, who, according to this school of thought, might be less concerned if they know that the repository is not 'for ever'. But there is more to it than this: in the beginning, retrievability was conceived as a temporary possibility, extended to a given period of time (decades, or a century). Nowadays, some look at retrievability as an operation to be allowed for an indefinite period of time, on the ground that *future generations may not agree upon geological disposal.*

One can easily imagine what this indefinitely 'open' solution may involve, not so much in terms of mining technique and operation but of costs, which would become infinite.

A clarification is due here, not of minor importance indeed: both additional barriers and retrievability, either temporary or indefinite, do not stem from the public's requests or from its representatives, nor from the so-called stakeholders, in other words from those people whose consensus must be achieved - a request which could perhaps justify them. These requirements originated instead within that technical and scientific community now envisaging an 'advanced democracy'. A technical community which, always willing to anticipate people's wishes, could have other surprises in store for us.

We know that, in nuclear engineering, whenever man and the environment's safety is really involved, as in nuclear reactor or reprocessing plant design and operation, the so-called redundancy criterion is applied. This means that, for equipment whose operation is essential for safety, a spare system is installed, that enters operation in case of a failure. Examples are the cooling water pumps in a reactor, or physical barriers, which, we should remember, had been provided for the Three Mile Island reactor but not for Chernobyl.

With a geological repository, a new concept of redundancy arose and spread, applied not to safety itself but to *its perception.* This means that redundancy, with the huge costs it implies, is not addressed at improving the safety level but at increasing people's confidence and at convincing them of the proposed solution's worth. Both more robust barriers and retrievability do not cancel, in fact - and we may wonder how they possibly could - the ultimate uncertainties which are associated with assessments extended to tens or hundreds of thousands of years. Indeed, to call things by their right name, in nuclear waste geological disposal, redundancy is an instrument of persuasion, i.e. of *marketing*. In the light of this, the demand for a broader consensus and the way of achieving it, does not really seem to be inspired, notwithstanding the fair language, by a higher democratic conception than the current one.

The problem, unfortunately, is that this useless redundancy takes us down a road of no return. As a matter of fact, once retrievability has been evoked, it will be quite hard, if not impossible, to do without it, even though it is not really necessary for safety. The public itself will claim it, since the technical experts themselves have considered it feasible and desirable. Actually, it has become almost a sort of dogma for most geological repository projects, and unfortunately tends to become so for short-lived waste repositories too, those presently being planned, where retrievability turns into the more radical concept of reversibility. More radical still, since here what should be retrievable is not just the waste, but the whole site, so the repository itself should be designed and constructed in a way that allows complete removal, in a more or less remote future.

In order to emphasize the need to gain public confidence in geological disposal and how to implement it, an alliance has taken place between nuclear waste experts and social science analysts, where the latter perform analyses that justify the positions and initiatives of the former. Among the most recurring analyses, being carried out over the last decade, are those based on the theory of risk perception and management, a subject which, in the NEA's publications about nuclear waste disposal, has taken the place previously occupied by scientific and technical issues.

Our social analysts have therefore made the great scientific discovery that nuclear energy generation in general, and radioactive waste in particular, are widely perceived, among all the activities and materials having potentially dangerous effects, as being those with the highest associated risk. Nothing new or sensational in this, of course. But according to most analysts, this attitude should be ascribed to two reasons: the historical circumstances under which nuclear energy appeared (the bomb and the connection between civilian and military uses) and the high level of secrecy which has always encompassed nuclear energy - the way, *from the top*, that decisions on nuclear energy have usually been taken.

How do these analyses differ from those of nuclear energy opponents? Cover-up's and accidents that have been kept concealed are not mentioned explicitly, but after all they may be considered included in "high secrecy level". By denouncing top-down decisions, the need for *democratic control of energy* is evoked, which was one of the virtues of the soft energies, as we may remember, and therefore considered as belonging to the Left.

We tried to explain in the previous chapters how the alleged faults of nuclear energy (civil-military mix-up, secrecy, cover-up, etc.) have only been propaganda weapons used by its opponents, and we also provided a casebook on how they have been manipulated, from the film *The China Syndrome* to *The Book of Greenpeace on the Nuclear Age*. It is surprising to see them proposed once again in NEA documents, more than twenty years later, as the causes of the public's negative perception of nuclear energy and, in particular, of radioactive waste. If one accepts them as the real causes, it means in fact recognizing that they are at least partially true, and this would indeed require a good dose of self-criticism - a real case of assuming a behaviour based on transparency, humility and dialogue, as evoked in the NEA's documents.

Sociology and anthropology experts, conceding that they have not been infected by ideological prejudice (even though, in reading their writings, some doubt arises), have only listened to the voices of nuclear opponents, without paying some attention to history. Had they done so, they would have found that a negative perception of nuclear energy did not exist among the public before the anti-nuclear movement entered into action, even though there had been Hiroshima and world peace was based on a nuclear deterrent, as discussed in a previous chapter. Looking at history, or simply at newspaper files, they would have discovered that, in creating a negative perception of nuclear energy, the main role was played by the mystification campaign launched by nuclear energy's ideological opponents, before and after Chernobyl.

It could be inferred that discussing the causes of the negative perception of nuclear energy and the difficulties that radioactive waste disposal encounters, only matters up to a certain point. One may say that, considering they exist, they must be faced up to and solved.

Indeed, this is the real problem: how to face and solve them depends upon their causes. If civil nuclear energy has been guilty of illicit management from its beginning, then clearly not only its image, but the cultural atmosphere surrounding it should be fully remodeled. It would then not be a question of inventing a special democracy, but simply of returning nuclear energy back to democracy, through a proper cultural restoration. And if the price to pay is to be a geological repository with many additional barriers, reversible and retrievable, to be considered like an open display piece over future centuries so as to convince the population, and perhaps even being fit to live in, then let us get on with it.

But if matters stand otherwise, if the negative perception of nuclear energy is the result of a quarter of a century of misinformation campaigns, including those surrounding the Chernobyl episode, then it's evident that the attitude to be adopted must be wholly different. The disinformation campaign, actually a slander campaign, succeeded because of a lack of adequate opposition and reaction.

Failing to denounce resolutely the mystification promulgated by the antinuclear activists has caused negative, and sometimes disastrous, scenarios attributed to nuclear energy to prevail in the collective imagination and has caused misgivings and concerns to settle out in people's perceptions. Radioactive waste experts, in particular, have not proudly claimed the seriousness of the studies performed and the scientific certainties acquired, and have thus given, and continues to give, the impression that much must still be done.

If these hesitations, doubts and weak reactions to misinformation have made possible - or at least favoured - the spread of a worried and anxious climate around nuclear waste disposal, then the problem will not be solved with the strategy that is dear to the NEA. The strategy which, to make geological disposal acceptable, calls for working in the long term with great humility and looking *sceptically* at technical achievements, in order to regain people's trust in nuclear experts.

What must be done, much more simply than by sophisticated social analyses, is to convince the interested communities that the solutions envisaged for waste disposal are safe. How?

Through communication, essentially based on demonstrating to what extent a proposed solution intends to care about protecting health and the environment. People will ask questions, raise queries - either technical or not - require guarantees, claim to have a role in project planning and development, demand compensation for land restraints and inconveniences, but they will expect clear and simple answers and explanations from experts, who, and this is of fundamental importance, must be capable of providing them.

Whoever has experienced, at least once in their professional life, discussing some aspects of radioactive waste disposal with ordinary people, in occasional meetings or in official hearings, knows that receptive and interested ears are usually found among the public of today's democratic countries. Ordinary people are willing to understand simple explanations, and, above all, to respect, not distrust, technical experts. And when, during this communication activity, some start speaking - as we can be sure will happen - of a nuclear dump, or spreading fear and dread about what may happen tomorrow or in ten thousand years, they must be refuted using simple and logical explanations, that scientists and engineers can readily and quietly provide, and the public easily comprehend - but never avoiding calling things by their names, thus identifying lies when someone is lying, and nonsense when someone talks nonsense.

Furthermore, and a decisive point, *explanations given by experts must address comparisons with the environmental impact of other industrial activities*, a subject that nuclear experts have inexplicably ever been reluctant to tackle, as if it were improper or incorrect. Instead, such comparisons have to be taken up vigorously, because people must know what risks and pollution they live with day by day, without however inciting anyone to rebellion! No one, whether scientist or sociologist, has ever considered launching campaigns to ascertain if those risks are acceptable or not, are avoidable or not, and, finally, guessing at what future generations might think about them.

Without mentioning major issues, like alternative means of energy generation, or the chemical industry, or the Kyoto protocol, but by remaining much closer to our day by day habits, we might wonder, for instance, why society accepts, without complains and worry for the future generations, the risks associated with using cars. Even just in the Western countries, this involves hundred of thousands of deaths per year, from accidents alone, without considering the much greater number of 'statistical deaths' from the associated pollution. Instead, this same society is induced to worry about one potential cancer case among one million inhabitants due to a remote contamination possibility, which may occur ten or one hundred thousand years from now. And whether, rather than mobilizing sociologists and scientists over geological disposal of tens of thousands of cubic metres of waste, it would not worth doing so for, say, investigating and developing a new transport concept, aimed at saving hundreds of thousands of human lives.

***

If social analysts have the rather anti-nuclear roles and positions we have seen them playing on nuclear waste disposal, and recommend behaviour (like scientific scepticism, humbleness, etc.) as an instrument of social pedagogy, how can the position of the scientific experts who lend them a hand be explained?

The experts who contribute to formulating and propagating the strategy that we have called 'advanced democracy', which finds its chosen audience in the NEA's radioactive waste committees, can, as we have anticipated, be divided in two groups: those who use this strategy to gain time and those who adopt it to lose time. Both, in one way or another, take the politicians' needs into account. Perhaps, there are also some who truly believe in it, as we will see. To better understand these experts, it is convenient to look briefly at what is happening in the countries they represent, limiting our attention to France and Germany, which, for opposite reasons, provide good examples supportive of our argument, and to Sweden, which is also a distinctive case.

In France, which is commonly acknowledged to be the most advanced country with regard to the nuclear option, two repositories for short-lived nuclear waste have been constructed, the second of which became operational in 1992 and is expected to be sufficient for disposing of all the short-lived waste that has ever been produced by the country. The case of long-lived waste is quite different, for in 1991 a dedicated law was issued, establishing a route whereby, within fifteen years (i.e. by 2006), the government shall present the parliament with a proposal for constructing a high-level waste repository. Technical decisions are entrusted to a National Evaluation Committee, purposely designated, to which national and international experts in this field have been appointed.

In a sense, this law is clearly, even if not officially, directed to geological disposal, since it specifies how and by what means and procedures this solution is to be implemented, including the construction of two deep underground laboratories to assess possible geological formations, clay and granite bedrocks. The law, however, does not designate the geological repository as the identified solution, but it establishes three alternatives be evaluated: geological disposal, long term (virtually indefinite) storage in near surface facilities, and transmutation. The latter is a process enabling very long-lived radioisotopes (like plutonium and other similar elements) to be eliminated from radioactive waste, so rendering very long-term isolation (hundreds of thousands of years and more) unnecessary. Without entering here into scientific details, we can simply say that transmutation is a nuclear process whose implementation, should its feasibility ever be proved, will require such enormous nuclear installations and time extension as to make it a purely imaginary alternative.

Since the French law was really conceived for making possible and regulating geological disposal, the procedures for site selection and for interacting with the local populations have been quite carefully drawn up.

For selecting the underground laboratory sites (one of which would become the actual repository, should the experimental activity be performed successfully), the principle of preliminary consultations with the local communities of the sites under consideration has been established. The modes of this consultation were fixed in a subsequent decree, creating the *négociateur* figure, a person specially appointed by the government for presenting the projects to local communities and for receiving possible voluntary candidatures. The law also indicated the activities to be performed following the selection of two laboratory sites. The construction is to be authorized through administrative channels, after a preliminary consultation with local and regional authorities has been carried out, through a procedure known as *enquête publique*, applicable in France in cases of national interest installations and regulated by law. Moreover, on the sites of the underground laboratories, information and control committees will be set up, formed by local members of municipal and national assemblies, environmental associations, farmers' unions and professional associations.

It appears clear then, that in Descartes' country, a rather rational route has been conceived, harmonizing the needs of science and technology, in other words the reasons of experts, and those of the local population, whose right is to be informed and exercise some control over project development (it is well known that in France, the local administrations have no power to veto governmental decisions).

Thus, the crucial question of so-called social acceptance is treated and regulated in a way that certainly is not the 'advanced democracy' we discussed, but what we would call 'simple' democracy. The project development is clearly of a *top-down* type, insofar as the basic technical choices, from site selection to man-made and geological barriers, are entrusted to those technically able to make them, and not open for discussion with people during the *enquête publique* and afterwards. Instead, the local population is provided with all the explanations, guarantees and verifications it is entitled to, while the public at large is satisfied, as occurs in democracy, by the laws freely approved by the national parliament.

The Cartesian French law worked rather well until 1999, and even benefited from some voluntary candidatures, thanks to the intelligent action of the *négociateur* and also to the widespread occurrence of deep clay bedrocks, which is one of the geological formations being considered. This made it possible to narrow down the site search to those not having opposition by local communities, and one of these was finally selected, where an underground laboratory has been under construction since 2001.

In 1999, when paper investigations to identify a number of sites in granite bedrock were under way for the second underground laboratory, disturbers from various schools entered into action. They created alarm among the populations living in the areas under consideration, also because this time a *négociateur* wasn't yet at work. The protest demonstrations alarmed the government, leading prime minister Jospin, who was already looking to the important elections planned for spring 2002, to suspend all site investigations in 2000.

However, the French geological disposal programme wouldn't be seriously jeopardized, because one underground laboratory is already under construction, though the time schedule, namely the decisions expected by 2006 as the law requires, might be out of reach. For this reason a new law is in preparation, adjusting the date and, perhaps, the objectives. Anyway, a manifest halt has been perceived in official statements and positions, beginning with those taken in the NEA committee by representatives of government agencies. Here, the French have become convinced supporters of a variation of the strategy calling for societal involvement in decisions about geological disposal: before making any technical decisions, an *analyse sociale* has to be carried out. What does this signify?

Most probably, that between the true alternatives under consideration in France, geological disposal and long term storage, the politicians intend to discover which one would be less controversial and conflict generating, and thus capable of impairing consensus towards the government (whichever in power, because in France Right and Left do not have different positions about energy, and the two sides do not fight over such matters of national interest). Should it emerge, from the social analysis, that long term storage is not expected to raise opposition and social conflicts (and it surely would not, because a storage system could be installed at one of the existing nuclear sites), then it's possible that the government would direct the experts along this route. The fact that this is not a definitive solution, nor the safest one in the long term, and also that it is the one involving much greater costs on the long term for the nation, and consequently for taxpayers, probably would not be considered a critical factor. In any case, no one would draw attention to this point, either in technical or in political circles. After all, societal peace is priceless.

Whilst waiting to find out the least disputed solution, what is better than to cultivate the NEA's philosophy? After all, postponing the decision, even for many years, does not compromise what is really important for the government: continued operation of the power plants and production of electricity (which in France is almost totally of nuclear origin). For this, nothing more is required than to provide additional space for high level waste storage.

If the French are among those taking advantage of the NEA's ultra-democratic strategies, the Germans share and encourage them because they suit the governmental policy, established by a coalition of socialists and greens, clearly aimed at preventing the solution of the nuclear waste problem. It might be hard to believe that any government does not want to solve such a national problem, but in Germany it is actually the case.

It would take too long to relate all the past and present disputes about nuclear waste disposal in that country, where the most extreme and violence-prone antinuclear movement in the world exists, so we shall dwell only on the main points. Incidentally, one should reflect on the fact that Italy and Germany, where there are such rabid antinuclear activists, are also the two Western countries having experienced in the twentieth century illiberal governing systems, set up with the decisive assistance of violent minorities.

Germany has conducted pioneering studies on deep disposal of long-lived waste since the seventies. Since 1979 studies and experiments have been concentrated on salt bedrocks, and a salt mine has been excavated at Gorleben, in Lower Saxony, which is perhaps, technically and geologically, the best location imaginable worldwide for a radioactive waste repository.

As soon as the so-called red-green government took over, after winning the 1998 elections, it immediately dedicated its attention to nuclear energy, which was historically the greens' bugbear, having fought it relentlessly for thirty years. Some may recall the protest occupations of public buildings (memorably that of Cologne cathedral), up to the recent assaults on trains carrying nuclear fuel or conditioned waste coming back from foreign reprocessing plants. Once within government, in dark suits, the antinuclear greens could not, of course, attack nuclear energy with their usual methods.

After having initially tried to ban nuclear power plants, in 2000 they passed a new law on nuclear energy, postponing the end of nuclear power to when the existing power plants (there are 19, producing about one third of the country's electricity) will cease operation on account of their age. At least for the short term, economic arguments prevailed over ideological ones.

It is, however, on radioactive waste management that the German greens have left their real mark. The experimental activity at the Gorleben salt mine, which for years had been not only the venue for all the world's nuclear waste experts but also the preferred site for mob gatherings of anti-nuclear activists, has been suspended for ten years. On what grounds? Because a new investigation must be carried out over the whole of Germany in order to identify the most suitable geological formation for long-lived waste disposal ..... At the same time, it has been stated that several conditions are to be fulfilled, and some issues to be more closely evaluated, none of which had actually been neglected at Gorleben, including, it goes without saying, retrievability and more robust man-made barriers to reinforce the natural ones. Besides, obviously, the requirement that the search and selection for a repository site be carried out with transparency and participation.

In Germany then, where an excellent site is already available, almost geologically ideal indeed, in which for twenty years studies of uncommon scope and seriousness have been carried out, it was decided to start again from zero, and to restart by looking again for suitable geological formations.

It's almost a classic of the anti-nuclear movement attitude: *the problem of nuclear waste disposal must not be allowed to appear soluble.* Just as Bertoldo, in Boccaccio's famous *Decameron*, is not willing to find the tree from which he is to be hanged, likewise the German greens, should they continue to be in the government coalition, will never allow a site to be chosen for radioactive waste disposal (including short-lived waste disposal). Should they be excluded from government and return to opposition, and then to radical protest (with or without camouflage headgear), they will thus still have a good reason to ask for nuclear power plant closures and the end of an energy source that is incapable of managing its waste.

It is therefore evident that the German representatives in the NEA's Radioactive Waste Management Committee, directly representing, by the way, a Ministry of the federal government, are among the most convinced supporters of the procedures recommended there, calling for large participation, for a general agreement within society, and for whatever can assist with indefinite procrastination.

If achieving societal confidence is a strategy that the French and Germans consider useful, although for opposite reasons, the Swedes are the ones who actually invented it. Unlike the former, however, they are implementing it in a quite positive way, because they really intend to go ahead with geological disposal and, above all, because they can afford to do so.

To begin with, such a strategy was conceived and pursued by them just to select a site, not to convince society as a whole that geological disposal is a feasible and safe solution. The latter is taken for granted, and indeed they have a sound technical solution for the repository, accurately identified and undergoing careful tests, on which a vast amount of information is being given to all the local communities having suitable geological sites in their territory. Advanced democracy, as conceived in Sweden, lies in any site being considered, even for a preliminary study on paper, only if a local community agrees, and in really trying to obtain a sort of global consent. Since 2002, two sites are being actively investigated in geotechnical and ecological field programmes, since the local authorities and populations have granted a permit to do so. A possible final choice, all the more so, will have to get the approval of local communities, before being submitted to national decision makers.

The geological disposal concept and the technical solution envisaged for emplacement of waste in the repository, which is presently being tested in one of the largest underground laboratories of its kind, are not open for discussion or debate (at least till today, 2003), but simply illustrated and presented to the public through a sort of 'capillary' information programme.

This strategy, based on gaining a true social consensus for site selection, which no other country yet implements so meticulously, is however a luxury that the Swedes can grant themselves on account of the situation they enjoy with respect to nuclear energy, and their national character.

As we mentioned previously, in Sweden a referendum decided to phase-out nuclear power production by 2010. The deadline now considered is 2020, since the nuclear power plants run very well. Indeed, many people are even distressed at the idea of having to renounce nuclear energy after that date. Meanwhile, however, whilst waiting for that deadline, in terms of nuclear electricity produced *per person*, Sweden is by far the number one in the world, and, as its nuclear power accounts for about 50% of the global national power output, it is ranked among the very first places among countries using nuclear energy. Sweden is then, in a sense, in the enviable position of being engaged in nuclear energy and in all the associated activities without any serious opposition, since, many years ago, they gave the opponents the bone that they wanted, if we may say so.

Actually, Sweden does things which, elsewhere, would cause at least some problems. For instance, all nuclear transports take place - completely unchallenged - by ship (naturally they are ultra-safe, but we know that this is not what counts for the opponents). For instance, an impressive underground repository for short-lived waste has been constructed under the Baltic seabed. By the way, this underground repository, one of the favourite sites of the international "nuclear tourism", is expected to lie beneath dry land in a couple of thousand years, because the Baltic Sea is retreating owing to slow, post-glacial land rise in Scandinavia. But the repository's safety has been proved and, of course, everyone believes it.

This is once more a luxury that Sweden can afford. A calm and collected population, trusting its authorities and even its politicians, perhaps more so than anybody else does in Europe (although what we have said for Sweden is also good for Finland, a country where more or less the same things are being done about nuclear waste, though on a smaller scale - actually Finland is the first European country where the construction of a geological repository has been authorized by the national parliament). Clearly, in Sweden, a route envisaging an almost global involvement of the population in site selection for nuclear waste disposal is not only feasible; considering the boundary conditions, it would even be the option of first choice. Not only is the *audience* responsive, but there are no professional troublemakers (perhaps because their tempers conform to the latitude).

We are quite confident that the public involvement strategy will be successful in Sweden, also because the governing class itself has only a minimum dose of what the French call *ésprit florentin* (and others call, less nobly, cunning), an attitude that elsewhere has much more influence on political decisions.

The trouble with the Swedes is that they, through the NEA committees where they have a certain weight on account of their achievements on waste disposal, feel themselves committed to export the notion of societal involvement and social consensus.

Outside the Scandinavian context, and especially in those countries where public and political debates sometimes call to mind the immortal satire in a celebrated comedy of Aristophanes, this noble strategy can only bring paralysis. This is particularly true in those cases where the social consensus principle has actually been extended from site selection to the very concept of geological disposal. In some countries, like Canada, where, after all, both environment and people are not much different from Scandinavia, the idea has been stretched to its extreme consequences. An independent panel, expressly appointed in 1988 to assess the safety of man and the environment in the geological disposal project presented by the Canadian nuclear agency, spent ten years to come up with this verdict:

> From a technical perspective, the safety of the proposed concept has been adequately demonstrated [..] As it stands, the concept of deep geological disposal has not been demonstrated to have broad public support.

This incredible conclusion had the effect of postponing decisions for years, and in 2002 a law on the disposal of long-lived waste (that in Canada comprises spent fuel alone) was passed, in which this requirement for a wide social involvement is incorporated. This will end up heavily influencing the evolution, duration and cost of the project.

The strategy contemplated by the NEA does not just have followers and inventors; it also has some sincere supporters among technical experts who are by no means newcomers (whether or not they represent a national policy). They sympathize with the social analysts' positions, which, as we have shown above, quite often end up by coinciding with those of antinuclear opponents.

Can they still be considered victims of the Chernobyl effect? Can they, that is, be numbered among those who have not yet recovered from that sort of historical lesson, according to which nothing can be taken for granted, and that doubting is a moral duty? Are they, in other words, sincerely convinced that the safety of geological disposal is not yet proven?

We have seen that nuclear waste disposal is definitely not the proper subject over which to exercise this kind of doubts, which could be admitted where nuclear reactor safety is concerned. When, instead, some technical expert cultivates societal analyses as the method to make radioactive waste disposal feasible, we suspect, more than a crisis of conscience, a case of scientific and intellectual narcissism, and also an attitude of self-promotion.

Only an attitude of this kind - forgetting for a while those who simply want to spend or lose time - can lead to suggestions for a decisional procedure (quoted from a recent publication) such as:

> At any stage in deciding between the major alternatives – disposing of nuclear waste in a repository or storing the waste in an interim storage facility – society (i.e., the public and political authority) must compare two types of uncertainties and decide *which is more acceptable*, or whether a combination of the two is preferable.

Anyone can imagine what a decision process of this sort may involve and how it be might established, but above all it appears quite clear that what is considered fundamental is not *the level of safety, but the level of acceptability of a technical solution.* So that a simple question arises: what will these technical experts do, should they find out by evaluating what people like better, that the most accepted solution is the least safe for man and the environment? Would they cheat people, not telling them that what they like is not the best?

In a sense, these *border-line* experts (because they are unintentionally bordering on the anti-nuclear opposition area) can be considered as the most sophisticated products of the long campaign against nuclear energy, and of the emphasis consequently given to the social acceptance stance.

Why narcissism and self-promotion? Because it is difficult to admit that a nuclear engineer or scientist, educated and used to discuss technical solutions in terms of functions and safety analysis, can place at the heart of the matter a sociological parameter such as social acceptability, and do that not because it is something of great concern for the whole humanity's destiny, which would be quite understandable. Narcissism also means gazing admiringly at ones own scientific and technical knowledge, considering it out of the reach of, or non-communicable to others, and being concerned *not with trying to divulge it but only with the public's perception of it*, showing by this no openness towards the public.

Self-promoting attitudes can be seen among experts who want to be a thorn in the side of those who intend to take action on nuclear waste disposal, and they promote themselves by raising and cultivating doubts and playing the part of the people's paladins.

They will start working, we can be sure, on the American Yucca Mountain project, which has caused some in the NEA Radioactive Waste Committee to grumble, as we have recalled above. Approved by Congress in July 2002 and thereafter undergoing the long procedure for getting a construction license from the nuclear safety authority, the repository project is not liked by many of these followers of the "scientific scepticism" recommended in the NEA's documents. They do not like it on account of the ten thousand years instead of the hundred thousand or million years they would have preferred; they do not like it because they say people have not been listened to enough; they do not like it on account of the geological bedrock selected; or because they claim that waste shipment would disturb Nevada's citizens (as though disturbing those of Texas or Minnesota was more democratic, or as though waste could fly); they do not like it because many of them do not like George Bush.

The fact is that the USA, after what they now simply call *September eleven*, feel themselves, if not at war, certainly on the alert: this is why they decided to speed up the project, which will allow them, among others, to secure seventy thousand tons of spent fuel, now spread over the whole country. And if it is impossible to prove by calculations what is going to happen ten thousand years from now, it is not seen as an issue over which to split hairs.

All the more so, as in the USA a geological repository for plutonium contaminated waste produced in the government's facilities is already in operation. This repository, called WIPP (Waste Isolation Pilot Plant), has been constructed seven hundred meters below ground, in a very large salt basin in southern New Mexico. In operation since 2000, after a licensing procedure which lasted almost twenty five years, having also obtained the agreement and even the favour of local and State authorities, WIPP has its unfailing faultfinders. In extra-prudent Europe, some claim that not all accidental scenarios have been assessed, among which is the possible complete dissolution of WIPP salt because of infiltration of *salt brine* from below, after ten thousand years ….

If, then, there are some who take advantage of the 'advanced' democracy concept to postpone decisions they do not feel like making, others who use it to postpone decisions they do not want to take and others, we might say, who try to be smart, the outcome is the same: *a helping hand is given to those who maintain that the problem of nuclear waste has not been solved yet, and so, at this point, can never be solved.*

It is worth stressing the concept again. Nuclear waste disposal is the last front where ideological environmentalism, in its long struggle against nuclear energy, has established itself as the most favourable at this moment in history. After having given up Chernobyl and reactor safety, technical issues into which only true experts may venture, the anti nuclear opponents have identified the final destination of nuclear waste as a field where they are free to deal with the non-exact sciences they like to handle.

When technical experts, from different schools of thought, allow and even encourage the scientific and technical achievements be put in the shade, on the basis that societal consensus is needed and that even simple political democracy is not sufficient, they come to the aid of anti-nuclear environmentalism.

If technical experts do not proudly and publicly claim that science and technology can nowadays ensure the safe disposal of nuclear waste, they allow the incapability of solving this problem to hang, like Damocles' sword, over nuclear energy.

***

Before concluding this chapter, where we have talked about nuclear waste and democracy, a subject is worth mentioning which, even if not directly dealing with participation and consensus, is still bound to raise feasibility and wide-ranging political problems, whose solution is definitely a major historical and cultural challenge: the *international* geological repository.

Although long considered heretical, the concept of implementing just one repository (or more than one, depending on its possible capacity and location) to dispose of long lived waste produced in several countries, is slowly gaining ground, along with generating quite a lot of worry. Why heretical, and why the worry?

The international geological repository for nuclear waste is one of those topics where we can see again how hard it is to have rational or even simple common sense discussions when dealing with nuclear energy. From the safety and environmental protection standpoint, which should be the real priority, a common repository for long lived waste - just one, but this is true also for a repository at regional or continental level - would be an ideal solution, for at least two reasons.

First, by concentrating all radioactive material in one site, not only would a worldwide dissemination in several repositories be avoided - leaving future generations with a nuclear legacy made of "holes" throughout the planet filled with harmful material - but both the security surveillance and safety monitoring operations needed in the medium term would also be greatly simplified and enhanced. Besides, in terms of environmental impact considered on a global perspective, the associated risk would be lower if the same material is concentrated in a single site instead of being spread in more than one (not to say if dispersed in dozens of them). This can be shown by calculation, but it is also easily understood if reference is made to one of the accidental scenarios which have to be considered in repository safety analyses, as seen in the previous chapter. Inadvertent intrusion occurring in a remote future, for instance, is an event whose probability surely does increase with the number of repositories, though remaining extremely low. With increasing numbers of repositories, the probability of habitat changes which may adversely affect barrier performance increases as well, since not all repositories would have the same endurance to changes. The same can be said for any of the so-called "perturbed" scenarios, for instance those involving hydrogeology or seismic events. On the other hand, the long term behaviour of either natural or man-made barriers is not affected by the amount of nuclear waste they have to confine, so the environmental impact of a repository isn't more critical if a larger amount of material is disposed of.

But it is mainly a second reason that makes the international site solution ideal. There are areas of Earth having extremely favourable geological and demographic features for a long lived waste repository. Areas which are the most tectonically stable, the most remote, the most geographically simple, the most arid in the world. Areas which cannot be influenced by habitat modification, nor by erosion or glaciation or similar phenomena, where there are geological formations assuring the highest isolation from water, and where, in addition, there is essentially no movement of water or no water at all. Areas which have no function in the global ecological equilibrium, and where, finally, the nearest human being (and in some case the nearest animal) lives hundreds of kilometers away. Such areas exist in the heart of Australia, in Patagonia, in eastern Asia, in Africa, in Russia. An area could be selected among these for an international repository site that would allow the problems of transportation, local geology and mining engineering to be simplified and optimized.

Quite often we have been asked, during discussions or hearings with ordinary people about final solutions for nuclear waste, whether it would be possible to launch them into outer space. Well, disposing them in one of the aforementioned areas is *more* than sending them into space, since the operation would be simpler, more easily controlled and consequently safer for man and the environment, and the isolation no less definitive.

Although few doubt that, on the safety and environmental protection side, an international geological repository would be the best solution, this is so difficult a matter to handle that in some milieus, and the NEA is one of these, it is taboo.

And yet, there are few concepts like this over which an international organization on waste management should take greater notice and care. First of all, many countries, including some under development, have, or will have, only small amounts of long lived radioactive waste, coming from one or few commercial and research reactors, from medical research or industry. It is inconceivable that each country with such waste could undertake a geological repository programme, primarily from economic considerations (a deep geological repository will cost from one to three billion Euro, no matter how small is the volume of wastes to be disposed); besides, not every country would be able to find suitable sites nationally. For such countries, then, a shared geological repository is the only viable solution. But it would also be *environmentally* inconceivable, because there would be a real proliferation of deep nuclear repositories (by the way containing also fissile material), almost in every area on Earth.

As expected, the antinuclear activists have been the first to denounce the international repository as heretical - in other words not politically correct. More than to the concept itself, their opposition has to be related to nuclear waste transportation, which they have always fought. They are indeed accustomed to oppose any shipment of nuclear material, and not on account of its possible risks, which are very low as everyone knows, themselves included. As a matter of fact, an assault on a train transporting nuclear waste, or Greenpeace boats playing at pirates attacking ships transporting spent fuel on the high seas, are activities giving a visibility that calculations on environmental impact or risk assessment will never achieve. So, the antinuclear activists oppose nuclear transportation, we would say as a rule, even though for them a nuclear shipment is an occasion for getting on television. But there is another argument leading them to deploy against international geological repositories. As we have seen, since the beginning, the antinuclear movement was marked by signs of antibourgeois and anti-Occidental ideology, and one of these is Third Worldism. Now it happens that some of those remote and wild aforementioned areas, actually the most suitable, lie in developing or poor countries, except Australia. We can imagine what could be the battle cry of antinuclear activists, should one of these countries – say one in Africa – be considered for hosting an international geological repository: the rich and colonialist West wants to dump its rubbish to the detriment of these poor countries' safety and health (or, in the case of Australia, they would claim that Aboriginal natural habitat is being violated).

Naturally, a shared repository could be one of those planned and under design in the industrial western countries, even though it wouldn't be at a site having such favourable conditions as one in earth's remote areas. But this is exactly what it is troubling, so to speak, some western countries' sleep. This is not as a result of any form of ecological nationalism; true experts are actually well aware that an international geological repository in France, Germany or the United Kingdom – for instance a regional one, housing all long lived waste from the European Union – would by no means make more serious the technical and environmental problems encountered for implementing a purely national repository. But some of these countries are pursuing a national geological disposal programme, and particularly, as we have seen, they are strongly committed, in order to select a site, to gaining the social consensus of local communities. What they fear is that, by raising the question of an international repository, even just by discussing the issue in a scientific committee or seminar, those communities might believe – or suspect – that *their* repository could become a shared one, and receive waste from the rest of the world. This, of course, could disturb the acceptance process under way.

It is not convincing, however, when those most opposed to the international geological repository notion, intransigent to such an extent that they refuse to talk about it, are those countries – like Sweden, France, the United Kingdom – that now preach the need for the global involvement of society in decisions about geological repositories. The same countries, in short, that claim today that a real social pedagogy process has to be implemented and ethical and moral values promoted in order to advance with nuclear waste disposal and site selection; a task, as we recalled above, worth undertaking for the far more important rationale of addressing real human needs.

Well, we believe that the international repository for long lived waste might be one of these prospects that really addresses the needs of people, globally and regionally. For the benefits it implies for the international community (we would say for mankind) and for global environment protection, promoting this concept would indeed deserve such extraordinary procedures, behavioural features and ethical considerations that the NEA recommends in its technical documents for a national geological repository. It is worth mentioning these again, to see whether we would then see them in a different light:

> A general agreement and confidence in society about the ethical, economic and political aspects of the geological disposal solution [..]
>
> A wide confidence basis on the practicability and long term safety of geological disposal [..]
>
> A wide confidence on organizational structures, on the legal framework and on established licensing and review procedures.
>
> Behavioural features to include openness, transparency, honesty, willingness to be "stretched", freedom from arrogance, recognition of limits, active search for dialogue [..]

Even the sophisticated approach suggested by those narcissists we mentioned above – the so-called staged repository – would be worth considering for such a task as an international geological repository. Instead, the NEA is paradoxically one of those international milieus where it is allowed to talk about an international geological repository only during the coffee breaks of a nuclear waste experts' meeting.

Of course, the NEA (namely the OECD) isn't the organization entitled to promote international geological disposal, though it has the technical qualification to support the concept at least. The natural place to deal with this matter is the United Nations, through the International Atomic Energy Agency. The scale of this problem is such that only the United Nations could manage it, particularly if, within all possible solutions, the ideal one would be pursued – a geological repository in one of the earth's remote areas.

Actually, at the IAEA the international disposal issue is beginning to be openly discussed, mainly by safety and radioprotection experts, who by education and practice are the most conscious of the advantages of such an answer to the problem of long lived nuclear waste disposal. It is however being discussed with due caution, paying attention to the words used and being careful not to disturb anyone, either nuclear industrial countries pursuing their national programmes, and having influence in the IAEA, or developing countries, which possess suitable areas and also constitute an influential group at the IAEA.

If the same rational attitude prevailed on nuclear energy as on the governance of so many other human activities, including some that are much more dangerous for man and the environment, we wouldn't be obliged to talk of a shared geological repository as a subject to be handled with extreme caution.

Few know that, in Germany, there are five salt mines, of the same kind as those considered for nuclear waste, where chemical toxic wastes coming from various European countries are disposed of. These materials are being shipped almost daily to the repositories, completely unchallenged and with no risk for the environment and human health (so we imagine, since we do not have direct knowledge of that) and, what's more, crossing several national borders. And this happens in Germany, where to transport vitrified nuclear waste and spent fuel by train, an army of policemen usually has to be mobilized (it's not hyperbole: in 2002 thirty thousand policemen had to intervene to protect a nuclear transport).

We can be sure that, for those five salt mines, absolutely nothing has been done that can be compared with twenty years of assessment and qualification of the Gorleben salt mine, former candidate to became the German disposal repository of nuclear long lived wastes, which are by no means more harmful than toxic waste. We can be sure as well that for emplacing these wastes no one has never envisaged a multiple barrier system nor the involvement of stakeholders or consideration of future generations' likely requirements. And finally, we know for sure that it never crossed anyone's mind to predict what could occur to disposed toxic waste and to the host salt formation in ten thousand or a hundred thousand years, nor to speak of retrieving them.

Such contradictions, a triumph of irrationality indeed, must be highlighted and explained to ordinary people and to those newcomers to the waste management arena: *the stakeholders*. These are information and explanations which can make people understand the facts and the needs of environmental policy, and so help to dispel the mystifications of the antinuclear activists. Insofar as it contributes to building up correct knowledge and to allowing people to free themselves from the influence of ideological opponents, this kind of information is not the occasion of environmental disputes, but an instrument of democracy – of normal and current democracy, not the 'advanced democracy' that we have met with in this closing chapter.

Fig.1: The final repository for low active nuclear waste of *Aube*, France.

Fig. 2: The final repository for low active nuclear waste of *El Cabril*, Spain.

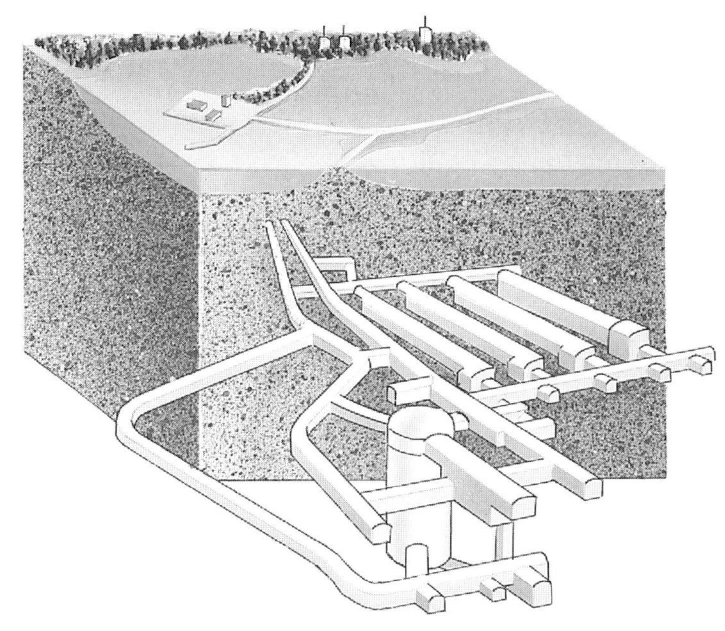

Fig. 3: A schematic view of the underground repository for low active nuclear waste of Forsmark,Sweden.

Fig. 4: Operation at Forsmark repository: handling of nuclear waste packages.

Fig. 5: The site of *Gorleben* salt mine, being considered for long lived nuclear waste, Lower Saxony,Germany.

Fig. 6: A tunnel of the Gorleben salt mine.

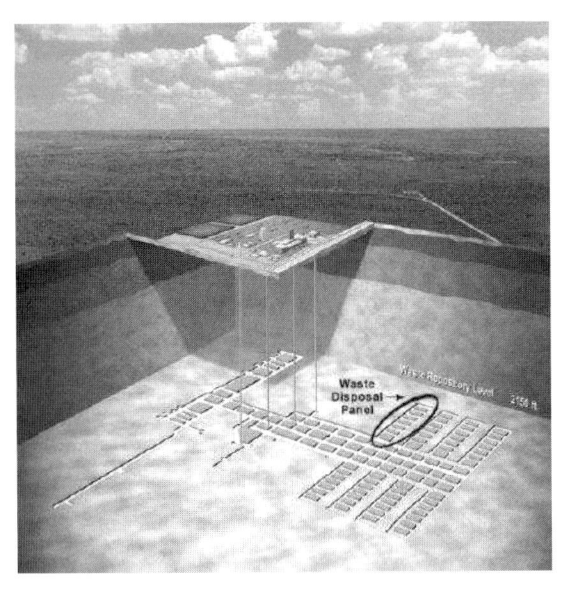

Fig. 7: The geological repository for long lived nuclear waste *WIPP*, New Mexico, USA.

Printing: Mercedes-Druck, Berlin
Binding: Stein+Lehmann, Berlin